Solids and Surfaces: A Chemist's View of Bonding in Extended Structures

Solids and Surfaces: A Chemist's View of Bonding in Extended Structures

Roald Hoffmann

Roald Hoffmann
Cornell University
Department of Chemistry
Baker Laboratory
Ithaca, New York 14853-1301

Library of Congress Cataloging-in-Publication Data

Hoffmann, Roald.
  Hoffmann. Solids and Surfaces: A Chemist's View on Bonding in Extended
Structures
  p.   cm.
  Bibliography: p.
  Includes index.
  ISBN 0-89573-709-4
  1. Chemical bonds.   2. Surface chemistry.   3. Solid state chemistry.
I. Title.
QD471.H83   1988
541.2'24--dc 19   88-14288
                                                                      CIP

British Library Cataloging in Publication Data

Hoffmann, Roald.
  Solids and Surfaces: A Chemist's View on Bonding in Extended Structures.
  1. Solids. Surfaces. Physical properties
  I. Title
  530.4'1

  ISBN 0-89573-709-4 US.

©1988 VCH Publishers, Inc.

Printed in the United States of America.

ISBN 0-89573-709-4 VCH Publishers
ISBN 3-527-26905-3 VCH Verlagsgesellschaft

Distributed in North America by: Distributed worldwide by:

VCH Publishers, Inc.   VCH Verlagsgesellschaft mbH
220 East 23rd St.   P.O. Box 1260/1280
Suite 909   D-6940 Weinheim
New York, New York 10010   Federal Republic of Germany

COVER: **Edge of Dunes** by Vivian Torrence

for
Earl Muetterties
and
Mike Sienko

Contents

Preface and Acknowledgments

The material in this book has been published in two articles in *Angewandte Chemie* and *Reviews of Modern Physics*, and I express my gratitude to the editors of these journals for their encouragement and assistance. The construction of the book based on those articles was suggested by my friend M. V. Basilevsky.

My graduate students, postdoctoral associates, and senior visitors to the group are responsible for both teaching me solid state physics and implementing the algorithms and computer programs that have made this work possible. While in my usual way I've suppressed the computations in favor of explanations, little understanding would have come without those computations. An early contribution to our work was made by Chien-Chuen Wan, but the real computational and interpretational advances came through the work of Myung-Hwan Whangbo, Charles Wilker, Miklos Kertesz, Tim Hughbanks, Sunil Wijeyesekera, and Chong Zheng. This book very much reflects their ingenuity and perseverance. Several crucial ideas were borrowed early on from Jeremy Burdett, such as using special k-point sets for properties.

Al Anderson was instrumental in getting me started in thinking about applying extended Hückel calculations to surfaces. A coupling of the band approach to an interaction diagram and frontier orbital way of thinking evolved from the study Jean-Yves Saillard carried out of molecular and surface C–H activation. We learned a lot together. A subsequent collaboration with Jérome Silvestre helped to focus many of the ideas in this book. Important contributions were also made by Christian Minot, Dennis Underwood, Shen-shu Sung, Georges Trinquier, Santiago Alvarez, Joel Bernstein, Yitzhak Apeloig, Daniel Zeroka, Douglas Keszler, William Bleam, Ralph Wheeler, Marja Zonnevylle, Susan Jansen, Wolfgang Tremel, Dragan Vučković, and Jing Li.

An important factor in the early stages of this work was my renewed collaboration with R. B. Woodward, prompted by our joint interest in organic conductors. Our collaboration was unfortunately cut short by his death in 1979. Thor Rhodin was mainly responsible for introducing me to the riches of surface chemistry and physics, and I am grateful to him and his students. It was always instructive to try to provoke John Wilkins.

Over the years my research has been steadily supported by the National Science Foundation's Chemistry Division. I owe Bill Cramer and his fellow program directors thanks for their continued support. A special role in my group's research on extended structures was played by the Materials Science Center (MSC) at Cornell University, supported by the Materials Research Division of the National Science Foundation. MSC furnished an interdisciplinary setting, which facilitated an interaction among researchers in the surface science and solid state areas that was very effective in introducing a novice to the important work in the field. I am grateful to Robert E. Hughes, Herbert H. Johnson, and Robert H. Silsbee, the MSC directors, for providing that supporting structure. In the last five years my surface-related research has been generously supported by the Office of Naval Research. That support is in the form of a joint research program with John Wilkins.

One reason it is easy to cross disciplines at Cornell is the existence of the Physical Sciences Library, with its broad coverage of chemistry and physics. I would like to thank Ellen Thomas and her staff for her contributions in that regard. Our drawings, a critical part of the way our research is presented, have been beautifully prepared over the years by Jane Jorgensen and Elisabeth Fields. I'd like to thank Eleanor Stagg, Linda Kapitany, and Lorraine Seager for their typing and secretarial assistance.

This manuscript was written while I held the Tage Erlander Professorship of the Swedish Science Research Council, NFR. The hospitality of Professor Per Siegbahn and the staff of the Institute of Theoretical Physics of the University of Stockholm and of Professor Sten Andersson and his crew at the Department of Inorganic Chemistry at the Technical University of Lund is gratefully acknowledged.

Finally, this book is dedicated to two men, colleagues of mine at Cornell in their time. They are no longer with us. Earl Muetterties played an important role in introducing me to inorganic and organometallic chemistry. Our interest in surfaces grew together. Mike Sienko and his students offered gentle encouragement by showing us the interesting structures on which they worked; Mike also taught me something about the relationship between research and teaching. This book is for them—both Earl Muetterties and Mike Sienko—who were so important and dear to me.

INTRODUCTION

Macromolecules extended in one, two, and three dimensions, of biological/natural or synthetic origin, fill the world around us. Metals, alloys, and composites, be they copper or bronze or ceramic, have played a pivotal and a shaping role in our culture. Mineral structures form the base of the paint that colors our walls and the glass through which we look at the outside world. Organic polymers, natural or synthetic, clothe us. New materials—inorganic superconductors, conducting organic polymers—exhibiting unusual electric and magnetic properties, promise to shape the technology of the future. Solid state chemistry is important, alive, and growing.[1]

So is surface science. A surface—be it of metal, an ionic or covalent solid, a semiconductor—is a form of matter with its own chemistry. In its structure and reactivity, it will bear resemblance to other forms of matter: bulk, discrete molecules in the gas phase and various aggregated states in solution. And it will have differences. Just as it is important to find the similarities, it is also important to note the differences. The similarities connect the chemistry of surfaces to the rest of chemistry, but the differences make life interesting (and make surfaces economically useful).

Experimental surface science is a meeting ground of chemistry, physics, and engineering.[2] New spectroscopies have given us a wealth of information, be it sometimes fragmentary, on the ways that atoms and molecules interact with surfaces. The tools may come from physics, but the questions that are asked are very chemical, e.g., what is the structure and reactivity of surfaces by themselves, and of surfaces with molecules on them?

The special economic role of metal and oxide *surfaces* in heterogeneous catalysis has provided a lot of the driving force behind current surface chemistry and physics. We always knew that the chemistry took place at the surface. But it is only today that we are discovering the basic mechanistic steps in heterogeneous catalysis. It's an exciting time; how wonderful to learn precisely how Döbereiner's lamp and the Haber process work!

What is most interesting about many of the new solid state materials are their electrical and magnetic properties. Chemists have to learn to measure these properties, not only to make the new materials and determine their structures. The history of the compounds that are at the center of today's exciting developments in high-temperature superconductivity makes this point very well. Chemists must be able to reason intelligently about the electronic structure of the compounds they make in order to understand how these properties and structures may be tuned. In a similar way, the study of surfaces must perforce involve a knowledge of the electronic structure of these extended forms of matter. This leads to the problem that learning the language necessary for addressing these problems, the language of solid state physics and band theory, is generally not part of the chemist's education. It should be, and the primary goal of this book is to teach chemists that language. I will show that it is not only easy, but that in many ways it includes concepts from molecular orbital theory that are very familiar to chemists.

I suspect that physicists don't think that chemists have much to tell them about bonding in the solid state. I would disagree. Chemists have built up a great deal of understanding, in the intuitive language of simple covalent or ionic bonding, of the structure of solids and surfaces. The chemist's viewpoint is often local. Chemists are especially good at seeing bonds or clusters, and their literature and memory are particularly well developed, so that one can immediately think of a hundred structures or molecules related to the compound under study. From empirical experience and some simple theory, chemists have gained much intuitive knowledge of the what, how, and why of molecules holding together. To put it as provocatively as I can, our physicist friends sometimes know better than we how to calculate the electronic structure of a molecule or solid, but often they do not *understand* it as well as we do, with all the epistemological complexity of meaning that "understanding" can involve.

Chemists need not enter into a dialogue with physicists with any inferiority feelings at all; the experience of molecular chemistry is tremendously useful in interpreting complex electronic structure. (Another reason not to feel inferior: until you synthesize that molecule, no one can study its properties! The synthetic chemist is very much in control.) This is not to say that it will not take some effort to overcome the skepticism of physicists regarding the likelihood that chemists can teach them something about bonding. I do want to mention here the work of several individuals in the physics community who have shown an unusual sensitivity to chemistry and chemical ways of thinking: Jacques Friedel, Walter A. Harrison, Volker Heine, James C. Phillips, Ole Krogh Andersen, and David Bullett. Their papers are always worth reading because of their attempt to build bridges between chemistry and physics.

I have one further comment before we begin. Another important interface is that between solid state chemistry, often inorganic, and molecular chemistry, both organic and inorganic. With one exception, the theoretical concepts that have served solid state chemists well have not been "molecular." At the risk of oversimplification, the most important of these concepts has been the idea that there are ions (electrostatic forces, Madelung energies) and that these ions have a certain size (ionic radii, packing considerations). This simple notion has been applied by solid state chemists even in cases of substantial covalency. What can be wrong with an idea that works, and that explains structure and properties? What is wrong, or can be wrong, is that application of such concepts may draw that field, that group of scientists, away from the heart of chemistry. The heart of chemistry, let there be no doubt, is the molecule! My personal feeling is that if there is a choice among explanations in solid state chemistry, one must select the explanation which permits a connection between the structure at hand and some discrete molecule, organic or inorganic. Making connections has inherent scientific value. It also makes "political" sense. Again, to state it provocatively, many solid state chemists have isolated themselves (no wonder that their organic or even inorganic colleagues aren't interested in what they do) by choosing not to see bonds in their materials.

Which, of course, brings me to the exception—the marvelous and useful Zintl concept.[3] The simple notion, introduced by Zintl and popularized by Klemm, Busmann, Herbert Schäfer, and others, is that in some compounds $A_x B_y$, where A is very electropositive relative to a main group element B, one could just think, that's all, think that the A atoms transfer their electrons to the B atoms, which then use them to form bonds. This very simple idea, in my opinion, is the single most important theoretical concept (and how *not* very theoretical it is!) in solid state chemistry of this century. And it is important not just because it explains so much chemistry, but because it forges a link between solid state chemistry and organic, or main group, chemistry.

In this book I will teach chemists some of the language of bond theory. As many connections as possible will be drawn to traditional ways of thinking about chemical bonding. In particular we will find and describe the tools—densities of states, their decompositions, crystal orbital overlap populations—for moving back from the highly delocalized molecular orbitals of the solid to local, chemical actions. The approach will be simple; indeed, oversimplified in parts. Where detailed computational results are displayed, they will be of the extended Hückel type[4] or of its solid state analogue, the tight-binding method with overlap. I will try to show how a frontier orbital and interaction diagram picture may be applied to the solid state or to surface bonding. There will be many effects similar to what we know happens for molecules. And there will be some differences.

ORBITALS AND BANDS IN ONE DIMENSION

It's usually easier to work with small, simple things, and one-dimensional infinite systems are particularly easy to visualize.[5-8] Much of the physics of two- and three-dimensional solids is present in one dimension.

Let's begin with a chain of equally spaced H atoms, **1**, or the isomorphic π system of a non-bond-alternating, delocalized polyene **2**, stretched out for the moment. And we will progress to a stack of Pt(II) square planar complexes, **3**, $Pt(CN)_4^{2-}$ or a model PtH_4^{2-}.

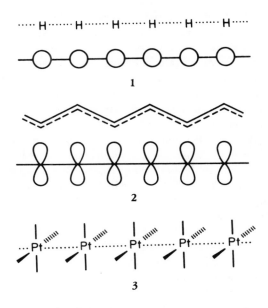

A digression here: every chemist would have an intuitive feeling for what that model chain of hydrogen atoms would do if released from the prison of its theoretical construction. At ambient pressure, it would form a chain of hydrogen molecules, **4**. This simple bond-forming process would be analyzed by the physicist (we will do it soon) by calculating a band for the equally spaced polymer, then seeing that it's subject to an instability, called a Peierls distortion. Other words around that characterization would be strong electron-phonon coupling, pairing distortion, or a $2k_F$ instability. And the physicist would come to the conclusion that the initially equally spaced H polymer would form a chain of hydrogen molecules. I mention this thought process here to make the point, which I will do repeatedly throughout this book, that the chemist's intuition is really excellent. But we must bring the languages of our sister sciences into correspondence. Incidentally, whether distortion **4** will take place at 2 megabars is not obvious and remains an open question.

Let's return to our chain of equally spaced H atoms. It turns out to be computationally convenient to think of that chain as an imperceptible bent segment of large ring (this is called applying cyclic boundary conditions).

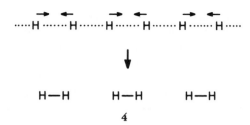

4

The orbitals of medium-sized rings on the way to that very large one are quite well known. They are shown in **5**. For a hydrogen molecule (or ethylene) there is bonding $\sigma_g(\pi)$ below an antibonding $\sigma_u^*(\pi^*)$. For cyclic H_3 or cyclopropenyl we have one orbital below two degenerate ones; for cyclobutadiene the familiar one below two below one, and so on. Except for the lowest (and occasionally the highest) level, the orbitals come in degenerate pairs. The number of nodes increases as one rises in energy. We'd expect the same for an infinite polymer—the lowest level nodeless, the highest with the maximum number of nodes. In between the levels should come in pairs, with a growing number of nodes. The chemist's representation of the band for the polymer is given at right in **5**.

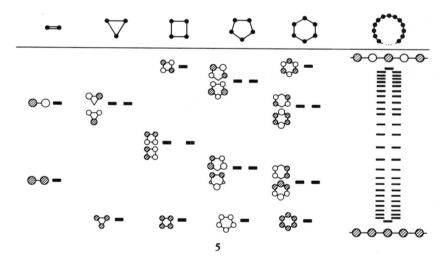

5

BLOCH FUNCTIONS, *k*, BAND STRUCTURES

There is a better way to write out all these orbitals by making use of the translational symmetry. If we have a lattice whose points are labeled by an index $n = 0, 1, 2, 3, 4 \cdots$ as shown in **6**, and if on each lattice point there is a basis function (a H 1s orbital), χ_0, χ_1, χ_2, etc., then the appropriate symmetry-adapted linear combinations (remember that translation is as good a symmetry operation as any other we know) are given in 6. Here a is the lattice spacing, the unit cell in one dimension, and k is an index that labels which irreducible representation of the translation group Ψ transforms as. We will see in a moment that k is much more, but for now k is just an index for an irreducible representation, just as a, e_1, e_2 in C_5 are labels.

$$\overset{\mid \leftarrow a \rightarrow \mid}{\underset{\substack{\chi_0 \quad \chi_1 \quad \chi_2 \quad \chi_3 \quad \chi_4}}{n = \; 0 \quad\;\; 1 \quad\;\; 2 \quad\;\; 3 \quad\;\; 4 \cdots}}$$

$$\psi_k = \sum_n e^{ikna} \; \chi_n$$

6

In the solid state physics trade, the process of symmetry adaptation is called "forming Bloch functions."[6,8-11] To reassure chemists that one is getting what one expects from 6, let's see what combinations are generated for two specific values of k: 0 and π/a. This is carried out in 7.

$$k = 0 \qquad \psi_0 = \sum_n e^0 \, \chi_n = \sum_n \chi_n$$

$$= \chi_0 + \chi_1 + \chi_2 + \chi_3 + \cdots$$

$$k = \frac{\pi}{a} \qquad \psi_{\frac{\pi}{a}} = \sum_n e^{\pi i n} \, \chi_n = \sum_n (-1)^n \, \chi_n$$

$$= \chi_0 - \chi_1 + \chi_2 - \chi_3 + \cdots$$

7

Referring back to 5, we see that the wave function corresponding to $k = 0$ is the most bonding one, the one for $k = \pi/a$ the top of the band. For other values of k we get a neat description of the other levels in the band. So k counts nodes as well. The larger the absolute value of k, the more nodes one has in the wave function. But one has to be careful—there is a range of k and if one goes outside of it, one doesn't get a new wave function, but rather repeats an old one. The unique values of k are in the interval $-\pi/a \leq k < \pi/a$ or $|k| \leq \pi/a$. This is called the first Brillouin zone, the range of unique k.

How many values of k are there? As many as the number of translations in the crystal or, alternatively, as many as there are microscopic unit cells in the macroscopic crystal. So let us say Avogadro's number, give or take a few. There is an energy level for each value of k (actually a degenerate pair of levels for each pair of positive and negative k values. There is an easily proved theorem that $E(k) = E(-k)$. Most representations of $E(k)$ do not give the redundant $E(-k)$, but plot $E(|k|)$ and label it as $E(k)$). Also the allowed values of k are equally spaced in the space of k, which is called reciprocal or momentum space. The relationship between $k = 1/\lambda$ and momentum derives from the de Broglie relationship $\lambda = h/p$. Remarkably, k is not only a symmetry label and a node counter, but it is also a wave vector, and so measures momentum.

So what a chemist draws as a band in **5**, repeated at left in **8** (and the chemist tires and draws ~ 35 lines or just a block instead of Avogadro's number), the physicist will alternatively draw as an $E(k)$ vs. k diagram at right. Recall that k is quantized, and there is a finite but large number of levels in the diagram at right. The reason it looks continuous is that this is a fine dot matrix printer; there are Avogadro's number of points jammed in there, and so it's no wonder we see a line.

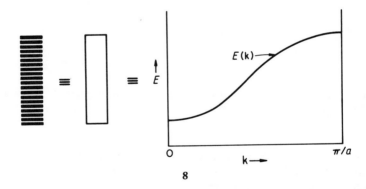

8

Graphs of $E(k)$ vs. k are called band structures. You can be sure that they can be much more complicated than this simple one. However, no matter how complicated they are, they can still be understood.

BAND WIDTH

One very important feature of a band is its *dispersion*, or *bandwidth*, the difference in energy between the highest and lowest levels in the band. What determines the width of bands? The same thing that determines the

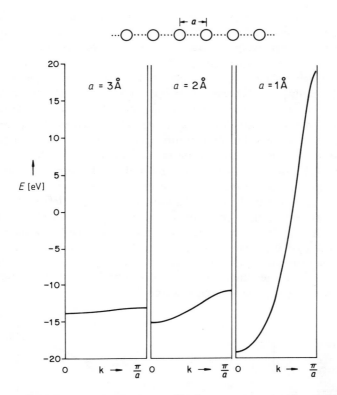

Figure 1 The band structure of a chain of hydrogen atoms spaced 3, 2, and 1 Å apart. The energy of an isolated H atom is − 13.6 eV.

splitting of levels in a dimer (ethylene or H²), namely, the overlap between the interacting orbitals (in the polymer the overlap is that between neighboring unit cells). The greater the overlap between neighbors, the greater the band width. Figure 1 illustrates this in detail for a chain of H atoms spaced 3, 2, and 1 Å apart. That the bands extend unsymmetrically around their "origin," the energy of a free H atom at − 13.6 eV, is a consequence of the inclusion of overlap in the calculations. For two levels, a dimer

$$E_\pm = \frac{H_{AA} \pm H_{AB}}{1 \pm S_{AB}}$$

The bonding E_+ combination is less stabilized than the antibonding one E_- is destabilized. There are nontrivial consequences in chemistry, for this is the source of four-electron repulsions and steric effects in one-electron theories.[11] A similar effect is responsible for the bands "spreading up" in Fig. 1.

SEE HOW THEY RUN

Another interesting feature of bands is how they "run." The lovely mathematical algorithm 6 applies in general; it does not say anything about the energy of the orbitals at the center of the zone ($k = 0$) relative to those at the edge ($k = \pi/a$). For a chain of H atoms it is clear that $E(k = 0) < E(k = \pi/a)$. But consider a chain of p functions, **9**. The same combinations as for the H case are given to us by the translational symmetry, but now it is clearly $k = 0$ that is high energy, the most antibonding way to put together a chain of p orbitals.

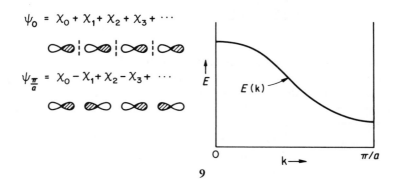

9

The band of s functions for the hydrogen chain "runs up," the band of p orbitals "runs down" (from zone center to zone edge). In general, it is the topology of orbital interactions that determines which way bands run.

Let me mention here an organic analogue to make us feel comfortable with this idea. Consider the through-space interaction of the three π bonds in **10** and **11**. The threefold symmetry of each molecule says that there must be an a and an e combination of the π bonds. And the theory of group representations gives us the symmetry-adapted linear combinations: for a, $\chi_1 + \chi_2 + \chi_3$; for e (one choice of an infinity), $\chi_1 - 2\chi_2 + \chi_3, \chi_1 - \chi_3$, where χ_1 is the π orbital of double bond 1, etc. But there is nothing in the group theory that tells us whether a is lower than e in energy. For that one needs chemistry or physics. It is easy to conclude from an evaluation of the orbital topologies that a is below e in **10**, but the reverse is true in **11**.

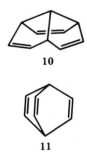

10

11

To summarize: *band width is set by inter-unit-cell overlap, and the way bands run is determined by the topology of that overlap.*

AN ECLIPSED STACK OF Pt(II) SQUARE PLANAR COMPLEXES

Let us test the knowledge we have acquired on an example slightly more complicated than a chain of hydrogen atoms. This is an eclipsed stack of square planar d^8 PtL_4 complexes, **12**. The normal platinocyanides [e.g., $K_2Pt(CN)_4$] indeed show such stacking in the solid state, at the relatively uninteresting $Pt \cdots Pt$ separation of ~ 3.3 Å. More exciting are the partially oxidized materials, such as $K_2Pt(CN)_4Cl_{0.3}$ and $K_2Pt(CN)_4(FHF)_{0.25}$. These are also stacked, but staggered, **13**, with a much shorter $Pt \cdots Pt$ contact of $2.7 \rightarrow 3.0$ Å. The Pt–Pt distance had been shown to be inversely related to the degree of oxidation of the material. [12]

$\longleftarrow a \longrightarrow$

12

$\longleftarrow 2a \longrightarrow$

13

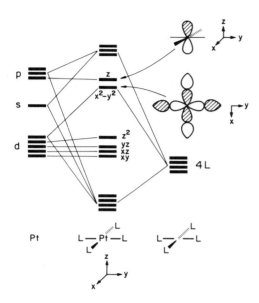

Figure 2 Molecular orbital derivation of the frontier orbitals of a square planar PtL$_4$ complex.

The real test of understanding is prediction. So let's try to predict the approximate band structure of **12** and **13** without a calculation, just using the general principles at hand. Let's not worry about the nature of the ligand; it is usually CN$^-$, but since it is only the square planar feature that is likely to be essential, let's imagine a theoretician's generic ligand H$^-$. We'll begin with **12** because its unit cell is the chemical PtL$_4$ unit, whereas the unit cell of **13** is doubled, (PtL$_4$)$_2$.

One always begins with the monomer. What are its frontier levels? The classical crystal field or molecular orbital picture of a square planar complex (Fig. 2) leads to a 4 below 1 splitting of the d block.[11] For 16 electrons we have z^2, xz, yz, and xy occupied and x^2–y^2 empty. Competing with the ligand field-destabilized x^2–y^2 orbital for being the lowest unoccupied molecular orbital (LUMO) of the molecule is the metal z. These two orbitals can be manipulated in understandable ways: π acceptors push z down, π donors push it up. Better σ donors push x^2–y^2 up.

We form the polymer. Each MO of the monomer generates a band. There may (will) be some further symmetry-conditioned mixing between orbitals of the same symmetry in the polymer (e.g., s and z and z^2 are of different symmetry in the monomer, but certain of their polymer molecular orbitals (MOs) are of the same symmetry). However, ignoring that secondary mixing and just developing a band from each monomer level independently represents a good start.

First, here is a chemist's judgment of the band widths that will develop: the bands that will arise from z^2 and z will be wide, those from xz, yz of medium width, those from x^2-y^2, xy narrow, as shown in **14**. This characterization follows from the realization that the first set of interactions (z, z^2) is σ type, and thus has a large overlap between unit cells. The xz, yz set has a medium π overlap, and the xy and x^2-y^2 orbitals (of course, the latter has a ligand admixture, but that doesn't change its symmetry) are δ.

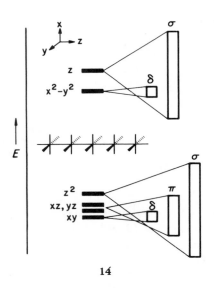

14

It is also easy to see how the bands run. Let's write out the Bloch functions at the zone center ($k = 0$) and zone edge ($k = \pi/a$). Only one of the π and δ functions is represented in **15**. The moment one writes these down, one sees that the z^2 and xy bands will run up from the zone center (the $k = 0$ combination is the most bonding) whereas the z and xz bands will run down (the $k = 0$ combination is the most antibonding).

The predicted band structure, merging considerations of band width and orbital topology, is that of **16**. To make a real estimate, one would need an actual calculation of the various overlaps, and these in turn would depend on the Pt· · ·Pt separation.

The actual band structure, as it emerges from an extended Hückel calculation at Pt–Pt = 3.0 Å, is shown in Fig. 3. It matches our expectations very precisely. There are, of course, bands below and above the frontier orbitals discussed; these are Pt–H σ and $\sigma*$ orbitals.

Here we can make a connection with molecular chemistry. The construction of **16**, an approximate band structure for a platinocyanide stack, involves no new physics, no new chemistry, no new mathematics

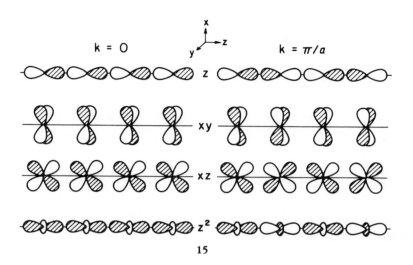

15

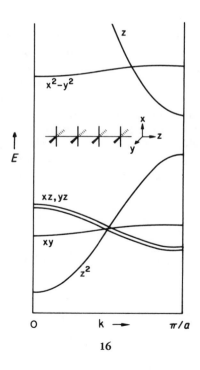

16

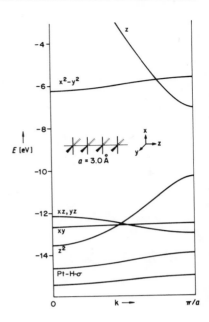

Figure 3 Computed band structure of an eclipsed PtH_4^{2-} stack, spaced at 3 Å. The orbital marked xz, yz is doubly degenerate.

beyond what every chemist already knows for one of the most beautiful ideas of modern chemistry: Cotton's construct of the metal–metal quadruple bond.[13] If we are asked to explain quadruple bonding, e.g., in $Re_2Cl_8^{2-}$, what we do is to draw **17**. We form bonding and antibonding combinations from the $z^2(\sigma)$, xz, $yz(\pi)$, and $x^2-y^2(\delta)$ frontier orbitals of each $ReCl_4^-$ fragment. And we split σ from σ^* by more than π from π^*, which in turn is split more than δ and δ^*. What goes on in the infinite solid is precisely the same thing. True, there are a few more levels, but the translational symmetry helps us out with that. It's really easy to write down the symmetry-adapted linear combinations, the Bloch functions.

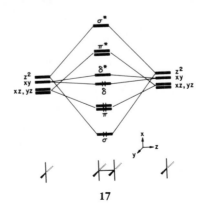

17

THE FERMI LEVEL

It's important to know how many electrons one has in one's molecule. Fe(II) has a different chemistry from Fe(III), and CR_3^+ carbocations are different from CR_3 radicals and CR_3^- anions. In the case of $Re_2Cl_8^{2-}$, the archetypical quadruple bond, we have formally Re(III), d^4, i.e., a total of eight electrons to put into the frontier orbitals of the dimer level scheme, **17**. They fill the σ, two π, and the δ level for the explicit quadruple bond. What about the $[PtH_4^{2-}]_\infty$ polymer **12**? Each monomer is d^8. If there are Avogadro's number of unit cells, there will be Avogadro's number of levels in each bond. And each level has a place for two electrons. So the first four bands are filled, the xy, xz, yz, z^2 bands. The Fermi level, the highest occupied molecular orbital (HOMO), is at the very top of the z^2 band. (Strictly speaking, there is another thermodynamic definition of the Fermi level, appropriate both to metals and semiconductors,[9] but here we will use the simple equivalence of the Fermi level with the HOMO.)

Is there a bond between platinums in this $[PtH_4^{2-}]_\infty$ polymer? We haven't yet introduced a formal description of the bonding properties of an orbital or a band, but a glance at **15** and **16** will show that the bottom of *each* band, be it made up of z^2, xz, yz, or xy, is bonding, and the top antibonding. Filling a band completely, just like filling bonding and antibonding orbitals in a dimer (think of He_2, and think of the sequence N_2, O_2, F_2, Ne_2), provides no net bonding. In fact, it gives net antibonding. So why does the unoxidized PtL_4 chain stack? It could be van der Waals attractions, not in our quantum chemistry at this primitive level. I think there is also a contribution of orbital interaction, i.e., real bonding, involving the mixing of the z^2 and z bands.[14] We will return to this soon.

The band structure gives a ready explanation for why the Pt$\cdots$Pt separation decreases on oxidation. A typical degree of oxidation is 0.3 electron per Pt.[12] These electrons must come from the top of the z^2 band. The degree of oxidation specifies that 15% of that band is empty. The states vacated are not innocent of bonding. They are strongly Pt–Pt σ antibonding. So it's no wonder that removing these electrons results in the formation of a partial Pt–Pt bond.

The oxidized material also has its Fermi level in a band, i.e., there is a zero band gap between filled and empty levels. The unoxidized platino-cyanides have a substantial gap—they are semiconductors or insulators. The oxidized materials are good low-dimensional conductors, which is a substantial part of what makes them interesting to physicists.[14]

In general, conductivity is not a simple phenomenon to explain, and there may be several mechanisms impeding the motion of electrons in a material.[9] A prerequisite for having a good electronic conductor is to have the Fermi level cut one or more bands (soon we will use the language of density of states to say this more precisely). One must beware, however, of (1) distortions that open up gaps at the Fermi level and (2) very narrow bands cut by the Fermi level because these will lead to localized states, not to good conductivity.[9]

MORE DIMENSIONS, AT LEAST TWO

Most materials are two- or three-dimensional, and while one dimension is fun, we must eventually leave it for higher dimensionality. Nothing much new happens, except that we must treat $\vec{k}$ as a vector, with components in reciprocal space, and the Brillouin zone is now a two- or three-dimensional area or volume.[9,15]

To introduce some of these ideas, let's begin with a square lattice, **18**, defined by the translation vectors $\vec{a}_1$ and $\vec{a}_2$. Suppose there is an H 1s orbital on each lattice site. It turns out that the Schrödinger equation in the crystal factors into separate wave equations along the x and y axes, each of them identical to the one-dimensional equation for a linear chain. There is a k_x and a k_y, the range of each is $0 \le |k_x|, |k_y| \le \pi/a \,(a = |\vec{a}_1| = |\vec{a}_2|)$. Some typical solutions are shown in **19**.

The construction of these is obvious. What the construction also

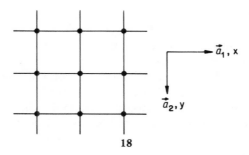

18

shows, very clearly, is the vector nature of k. Consider the $(k_x, k_y) = (\pi/2a, \pi/2a)$ and $(\pi/a, \pi/a)$ solutions. A look at them reveals that they are waves running along a direction that is the vector sum of k_x and k_y, i.e., on a diagonal. The wavelength is inversely proportional to the magnitude of that vector.

The space of k here is defined by two vectors $\vec{b}_1$ and $\vec{b}_2$, and the range

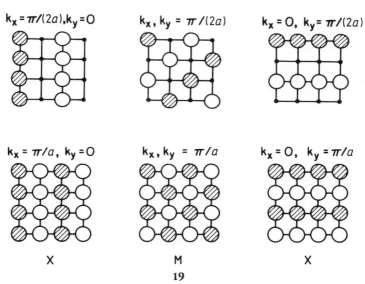

19

of allowed k, the Brillouin zone, is a square. Certain special values of k are given names: $\Gamma = (0, 0)$ is the zone center, $X = (\pi/a, 0) = (0, \pi/a)$, $M = (\pi/a, \pi/a)$. These are shown in **20**, and the specific solutions for Γ, X, and M were so labeled in **19**.

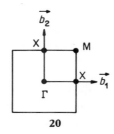

20

It is difficult to show the energy levels $E(\vec{k})$ for all $\vec{k}$. So what one typically does is to illustrate the evolution of E along certain lines in the Brillouin zone. Some obvious ones are $\Gamma \to X, \Gamma \to M, X \to M$. From **19** it is clear that M is the highest energy wave function, and that X is pretty much nonbonding, since it has as many bonding interactions (along y) as it does antibonding ones (along x). So we would expect the band structure to look like **21**. A computed band structure for a hydrogen lattice with $a = 2.0$ Å (Fig. 4) confirms our expectations.

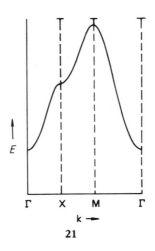

21

The chemist would expect the chessboard of H atoms to distort into one of H_2 molecules. (An interesting problem is how many different ways there are to accomplish this.)

Let's now put some p orbitals on the square lattice, with the direction perpendicular to the lattice taken as z. The p_z orbitals will be separated from p_y and p_x by their symmetry. Reflection in the plane of the lattice remains a good symmetry operation at all k. The $p_z(z)$ orbitals will give a band structure similar to that of the s orbital, since the topology of the interaction of these orbitals is similar. This is why in the one-dimensional case we could talk at one and the same time about chains of H atoms and polyenes.

The p_x, p_y (x, y) orbitals present a somewhat different problem. Shown below in **22** are the symmetry-adapted combinations of each at Γ, X, Y, and M. (Y is by symmetry equivalent to X; the difference is just in the propagation along x or y.) Each crystal orbital can be characterized by the

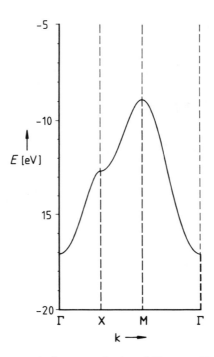

Figure 4 The band structure of a square lattice of H atoms, H–H separation 2.0 Å.

p,p σ or π bonding present. Thus at Γ the x and y combinations are σ antibonding and π bonding; at X they are σ and π bonding (one of them), and σ and π antibonding (the other). At M they are both σ bonding, π antibonding. It is also clear that the x, y combinations are degenerate at Γ and M (and, it turns out, along the line $\Gamma \to M$, but for that one needs a little group theory[15]) and nondegenerate at X and Y (and everywhere else in the Brillouin zone).

Putting in the estimate that σ bonding is more important than π bonding, one can order these special symmetry points of the Brillouin zone in energy and draw a qualitative band structure. This is Fig. 5. The actual appearance of any real band structure will depend on the lattice spacing. Band dispersions will increase with short contacts, and complications due to s, p mixing will arise. Roughly, however, *any* square lattice—be it the P net in GdPS,[16] a square overlayer of S atoms absorbed on Ni(100),[17] the oxygen and lead nets in litharge,[18] or a Si layer in BaPdSi$_3$[19]—will have these orbitals.

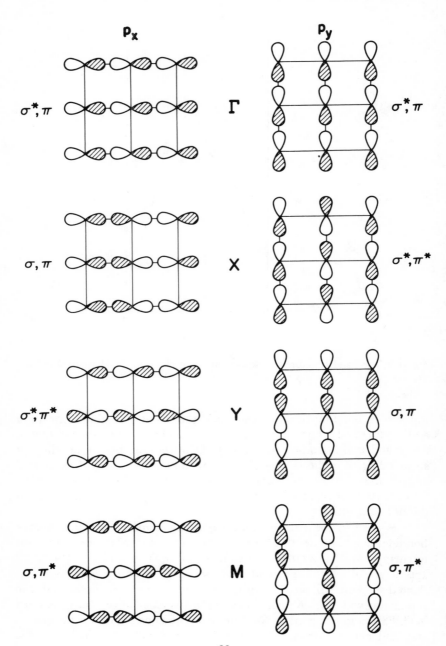

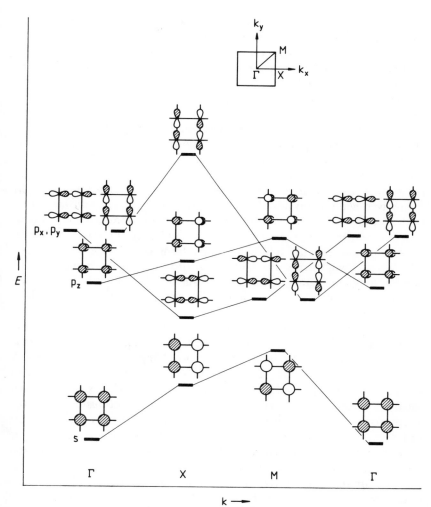

Figure 5 Schematic band structure of a planar square lattice of atoms bearing ns and np orbitals. The s and p levels have a large enough separation that the s and p band do not overlap.

SETTING UP A SURFACE PROBLEM

The strong incentive for moving to at least two dimensions is that obviously one needs this for studying surface-bonding problems. Let's begin to set these up. The kind of problems we want to investigate, for example, are how CO chemisorbs on Ni; how H_2 dissociates on a metal surface; how acetylene bonds to Pt(111) and then rearranges to vinylidene or ethylidyne; how surface carbide or sulfide affects the chemistry of CO; how CH_3 and CH_2 bind, migrate, and react on an iron surface. It makes sense to look first at structure and bonding in the stable or metastable waypoints, i.e., the chemisorbed species. Then one could proceed to construct potential energy surfaces for motion of chemisorbed species on the surface, and eventually for reactions.

The very language I have used here conceals a trap. It puts the burden of motion and reactive power on the chemisorbed molecules, and not on the surface, which might be thought of as passive, untouched. Of course, this can't be so. We know that exposed surfaces reconstruct, i.e., make adjustments in structure driven by their unsaturation.[20] They do so first by themselves, without any adsorbate. And they do it again, in a different way, in the presence of adsorbed molecules. The extent of reconstruction is great in semiconductors and extended molecules, and generally small in molecular crystals and metals. It can also vary from crystal face to face. The calculations I will discuss deal with metal surfaces. One is then reasonably safe (we hope) to assume minimal reconstruction. It will turn out, however, that the signs of eventual reconstruction are to be seen even in these calculations.

It might be mentioned here that reconstruction is not a phenomenon reserved for surfaces. In the most important development in theoretical inorganic chemistry in the 1970s, Wade[21a] and Mingos[21b] provided us with a set of skeletal electron pair counting rules. These rationalize the related geometries of borane and transition metal clusters. One aspect of their theory is that if the electron count increases or decreases from the appropriate one for the given polyhedral geometry, the cluster will adjust geometry—open a bond here, close one there—to compensate for the different electron count. Discrete molecular transition metal clusters and polyhedral boranes also reconstruct.

Returning to the surface, let's assume a specific surface plane cleaved out, frozen in geometry, from the bulk. That piece of solid is periodic in two dimensions, semi-infinite, and aperiodic in the direction perpendicular to the surface. Half of infinity is much more painful to deal with than all of infinity because translational symmetry is lost in that third dimension. And that symmetry is essential in simplifying the problem—one doesn't want to be diagonalizing matrices of the degree of Avogadro's number; with translational symmetry and the apparatus of the theory of group representations, one can reduce the problem to the size of the number of orbitals in the unit cell.

So one chooses a slab of finite depth. Diagram **23** shows a four-layer slab model of a (111) surface of an fcc metal, a typical close-packed hexagonal face. How thick should the slab be? Thick enough so that its

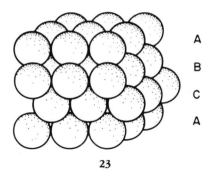

23

inner layers approach the electronic properties of the bulk, the outer layers those of the true surface. In practice, it is more often economics that dictates the typical choice of three or four layers.

Molecules are then brought up to this slab—not one molecule, for that would ruin the desirable two-dimensional symmetry, but an entire array or layer of molecules maintaining translational symmetry.[22] This immediately introduces two of the basic questions of surface chemistry: coverage and site preference. Diagram **24** shows a c(2 × 2) CO array on Ni(100), on-top adsorption, coverage = 1/2. Diagram **25** shows four possible ways of adsorbing acetylene in a coverage of 1/4 on top of Pt(111). The hatched area is the unit cell. The experimentally preferred mode is the threefold bridging one, **25c**. Many surface reactions are coverage-dependent.[2] And the position where a molecule sits on a surface, its orientation relative to the surface, is something one wants to know.

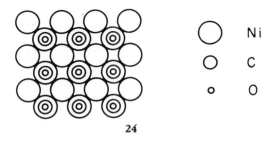

24

So we have a slab, three or four atoms thick, of a metal, and a monolayer of adsorbed molecules. Figure 6 shows what the band structure looks like for some CO monolayers, and Fig. 7 for a four-layer Ni(100) slab. The phenomenology of these band structures should be clear by now; it is

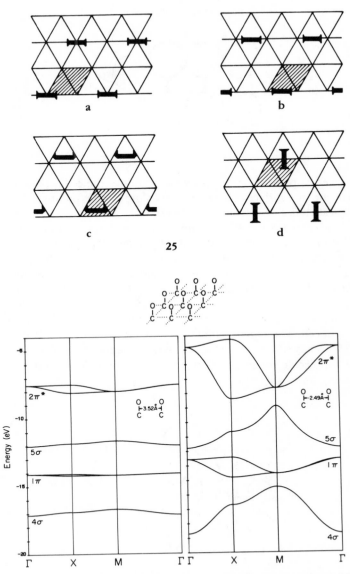

Figure 6 Band structures of square monolayers of CO at two separations: (a) left, 3.52 Å, (b) right, 2.49 Å. These would correspond to 1/2 and full coverage of a Ni(100) surface.

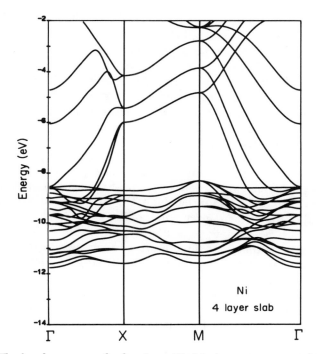

Figure 7 The band structure of a four-layer Ni slab that serves as a model for a Ni(100) surface. The flat bands are derived from Ni 3d; the more highly dispersed ones above these are 4s, 4p.

spelled out by the following:

(1) *What is being plotted:* E vs. $\vec{k}$. The lattice is two-dimensional. k is now a vector, varying within a two-dimensional Brillouin zone, $\vec{k} = (k_x, k_y)$. Some of the special points in this zone are given canonical names: Γ (the zone center) $= (0, 0)$; $X = (\pi/a, 0)$, $M = (\pi/a, \pi/a)$. What is being plotted is the variation of the energy along certain specific directions in reciprocal space connecting these points.

(2) *How many lines there are:* There are as many lines as there are orbitals in the unit cell. Each line is a band, generated by a single orbital in the unit cell. In the case of CO, there is one molecule per unit cell, and that molecule has well-known 4σ, 1π, 5σ, and $2\pi^*$ MOs. Each generates a band. In the case of the four-layer Ni slab, the unit cell has four Ni atoms. Each has five 3d, one 4s, and three 4p basis functions. We see some, but not all, of the many bands these orbitals generate in the energy window shown in Fig. 7.

(3) *Where (in energy) the bands are:* The bands spread out, more or less dispersed, around a "center of gravity." This is the energy of that

orbital in the unit cell that gives rise to the band. Therefore, 3d bands lie below 4s and 4p for Ni, and 5σ below $2\pi^*$ for CO.

(4) *Why some bands are steep, others flat:* This is because there is much inter-unit-cell overlap in one case, little in another. The CO monolayer bands in Fig. 6 are calculated at two different CO–CO spacings, corresponding to different coverages. It's no surprise that the bands are more dispersed when the COs are closer together. In the case of the Ni slab, the s, p bands are wider than the d bands, because the 3d orbitals are more contracted, less diffuse than the 4s, 4p.

(5) *Why the bands are the way they are:* They run up or down along certain directions in the Brillouin zone as a consequence of symmetry and the topology of orbital interaction. Note the phenomenological similarity of the behavior of the σ and π bands of CO in Fig. 6 to the schematic, anticipated course of the s and p bands of Fig. 5.

There are more details to be understood, of course. But, in general, these diagrams are complicated not because of any mysterious phenomenon but because of *richness*, the natural accumulation of understandable and understood components.

We still have the problem of how to talk about all these highly delocalized orbitals, and how to retrieve a local, chemical, or frontier orbital language in the solid state. There is a way.

DENSITY OF STATES

In the solid, or on a surface, both of which are just very large molecules, one has to deal with a very large number of levels or states. If there are n atomic orbitals (basis functions) in the unit cell, generating n molecular orbitals, and if in our macroscopic crystal there are N unit cells (N is a number that approaches Avogadro's number), then we will have Nn crystal levels. Many of these are occupied and, roughly speaking, they are jammed into the same energy interval in which we find the molecular or unit cell levels. In a discrete molecule we are able to single out one orbital or a small subgroup of orbitals as being the frontier, or valence orbitals of the molecules, responsible for its geometry, reactivity, etc. There is no way in the world that a single level among the myriad Nn orbitals of the crystal will have the power to direct a geometry or reactivity.

There is, however, a way to retrieve a frontier orbital language in the solid state. We cannot think about a single level, but perhaps we can talk about bunches of levels. There are many ways to group levels, but one pretty obvious way is to look at all the levels in a given energy interval. The density of states (DOS) is defined as follows:

$$DOS(E)dE = \text{number of levels between } E \text{ and } E + dE$$

For a simple band of a chain of hydrogen atoms, the DOS curve takes on the shape of **26**. Note that because the levels are equally spaced along the k axis and because the $E(k)$ curve, the band structure, has a simple cosine curve shape, there are more states in a given energy interval at the top and bottom of this band. In general, DOS(E) is proportional to the inverse of the slope of $E(k)$ vs. k, or, to say it in plain English, the flatter the band, the greater the density of states at that energy.

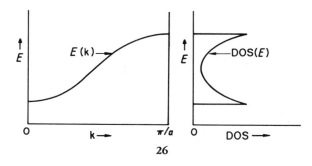

26

The shapes of DOS curves are predictable from the band structures. Figure 8 shows the DOS curve for the PtH_4^{2-} chain, Fig. 9 for a two-dimensional monolayer of CO. These could have been sketched from their respective band structures. In general, the detailed construction of these is a job best left to computers.

The DOS curve counts levels. The integral of DOS up to the Fermi level is the total number of occupied MOs. Multiplied by 2, it's the total number of electrons, so that the DOS curves plot the distribution of electrons in energy.

One important aspect of the DOS curves is that they represent a return from reciprocal space, the space of k, to real space. The DOS is an average over the Brillouin zone, i.e., over all k that might give molecular orbitals at the specified energy. The advantage here is largely psychological. If I may be permitted to generalize, I think chemists (with the exception of crystallographers) by and large feel themselves uncomfortable in reciprocal space. They'd rather return to, and think in, real space.

There is another aspect of the return to real space that is significant: *chemists can sketch the DOS of any material, approximately, intuitively.* All

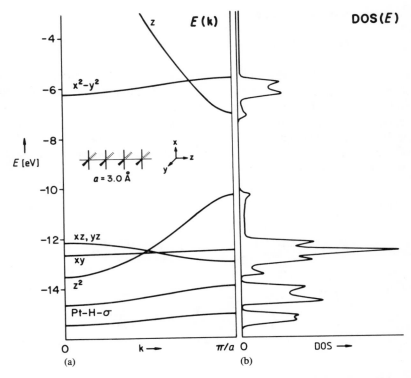

Figure 8 Band structure and density of states for an eclipsed PtH_4^{2-} stack. The DOS curves are broadened so that the two-peaked shape of the xy peak in the DOS is not resolved.

that's involved is a knowledge of the atoms, their approximate ionization potentials and electronegativities, and some judgment as to the extent of inter-unit-cell overlap (usually apparent from the structure).

Let's take the PtH_4^{2-} polymer as an example. The monomer units are clearly intact in the polymer. At intermediate monomer–monomer separations (e.g., 3 Å) the major inter-unit-cell overlap is between z^2 and z orbitals. Next is the xz, yz π-type overlap; all other interactions are likely to be small. Diagram **27** is a sketch of what we would expect. In **27** I haven't been careful to draw the integrated areas commensurate to the actual total number of states, nor have I put in the two-peaked nature of the DOS each level generates; all I want to do is to convey the rough spread of each band. Compare **27** to Fig. 8.

This was easy, because the polymer was built up of molecular monomer units. Let's try something inherently three-dimensional. The rutile structure of TiO_2 is a relatively common type. As **28** shows, the rutile

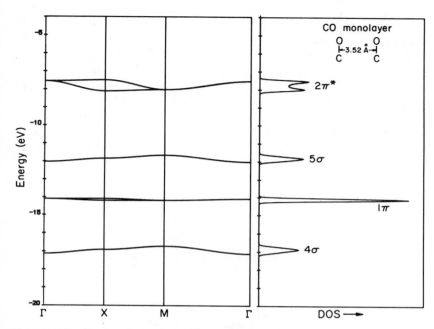

Figure 9 The density of states (right) corresponding to the band structure (left) of a square monolayer of CO's, 3.52 Å apart.

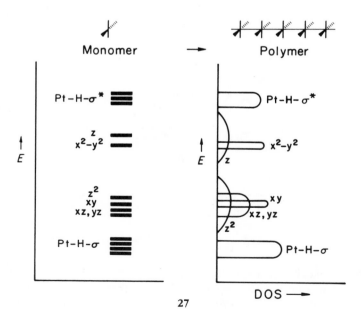

structure has a nice octahedral environment of each metal center, each ligand (e.g., O) bound to three metals. There are infinite chains of edge-sharing MO_6 octahedra running in one direction in the crystal, but the metal–metal separation is always relatively long.[23] There are no monomer units here, just an infinite assembly. Yet there are quite identifiable octahedral sites. At each, the metal d block must split into t_{2g} and e_g combinations, the classic three-below-two crystal field splitting. The only other thing we need is to realize that O has quite distinct 2s and 2p levels, and that there is no effective $O\cdots O$ or $Ti\cdots Ti$ interaction in this crystal. We expect something like **29**.

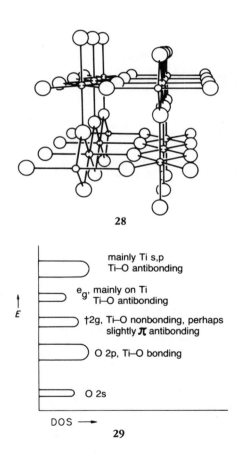

28

29

Note that the writing down of the approximate DOS curve *bypasses* the band structure calculation per se. Not that that band structure is very complicated; but it is three-dimensional, and our exercises so far have been

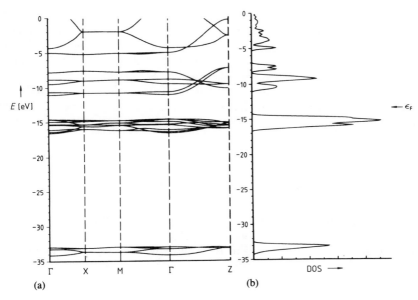

Figure 10 Band structure and density of states for rutile, TiO_2.

easy, in one or two dimensions. So the computed band structure in Fig. 10 will seem complex. The number is doubled (i.e., 12 O 2p, 6 t_{2g} bands), simply because the unit cell contains two formula units, $(TiO_2)_2$. There is not one reciprocal space variable, but several lines ($\Gamma \rightarrow X$, $X \rightarrow M$, etc.) that refer to directions in the three-dimensional Brillouin zone. If we glance at the DOS, we see that it does resemble the expectations of **21**. There are well-separated O 2s, O 2p, Ti t_{2g} and e_g bands.[23]

Would you like to try something a little (but not much) more challenging? Attempt to construct the DOS of the new superconductors based on the La_2CuO_4 and $YBa_2Cu_3O_7$ structures. And when you have done so and found that these should be conductors, reflect on how that doesn't allow you yet, did not allow anyone, to predict that compounds slightly off these stoichiometries would be remarkable superconductors.[24]

The chemist's ability to write down approximate DOS curves should not be slighted. It gives us tremendous power, qualitative understanding, and an obvious connection to local, chemical viewpoints such as the crystal or ligand field model. I want to mention here one solid state chemist, John B. Goodenough, who has shown over the years, and especially in his prescient book *Magnetism and Chemical Bonding*, just how good the chemist's approximate construction of band structures can be.[25]

However, in **27** and **29**, the qualitative DOS diagrams for PtH_4^{2-} and

TiO$_2$, there is much more than a guess at a DOS. There is a chemical characterization of the localization in real space of the states (are they on Pt? on H? on Ti? on O?) and a specification of their bonding properties (Pt–H bonding, antibonding, nonbonding, etc.). The chemist asks right away, where in space are the electrons? Where are the bonds? There must be a way that these inherently chemical, local questions can be answered, even if the crystal molecular orbitals, the Bloch functions, delocalize the electrons over the entire crystal.

WHERE ARE THE ELECTRONS?

One of the interesting tensions in chemistry is between the desire to assign electrons to specific centers, deriving from an atomic, electrostatic view of atoms in a molecule, and the knowledge that electrons are not as localized as we would like them to be. Let's take a two-center molecular orbital:

$$\Psi = c_1\chi_1 + c_2\chi_2$$

where χ_1 is on center 1 and χ_2 on center 2. Let's assume that centers 1 and 2 are not identical, and that χ_1 and χ_2 are normalized but not orthogonal. The distribution of an electron in this MO is given by $|\Psi|^2$. Ψ should be normalized, so

$$1 = \int |\Psi|^2 \, d\tau = \int |c_1\chi_1 + c_2\chi_2|^2 \, d\tau = c_1^2 + c_2^2 + 2c_1c_2S_{12}$$

where S_{12} is the overlap integral between χ_1 and χ_2. This is how one electron in Ψ is distributed. Now it's obvious that c_1^2 is to be assigned to center 1, c_2^2 to center 2. $2c_1c_2S_{12}$ is clearly a quantity that is associated with interaction. It's called the overlap population, and we will soon relate it to the bond order. But what are we to do if we persist in wanting to divide up the electron density between centers 1 and 2? We want all the parts to add up to 1, and $c_1^2 + c_2^2$ won't do. We must somehow assign the "overlap density" $2c_1c_2S_{12}$ to the two centers. Mulliken suggested (and that's why we call this a Mulliken population analysis[20]) a democratic solution, splitting $2c_1c_2S_{12}$ equally between centers 1 and 2. Thus center 1 is assigned $c_1^2 + c_1c_2S_{12}$, center 2 $c_2^2 + c_1c_2S_{12}$ and the sum is guaranteed to be 1. It should be realized that the Mulliken prescription for partitioning the overlap density, while uniquely defined, is quite arbitrary.

What a computer does is just a little more involved, since it sums these

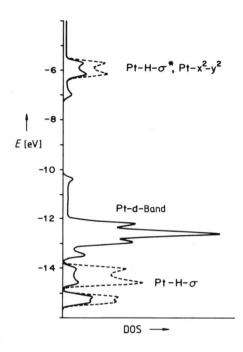

Figure 11 The solid line is the Pt contribution to the total DOS (dashed line) of an eclipsed PtH_4^{2-} stack. What is not on Pt is on the four H's.

contributions for each atomic orbital on a given center (there are several) over each occupied MO (there may be many). In the crystal, it does that sum for several k points in the Brillouin zone, and then returns to real space by averaging over these. The net result is a partitioning of the total DOS into contributions to it by either atoms or orbitals. We have also found very useful a decomposition of the DOS into contributions of fragment molecular orbitals (FMOs); i.e. the MOs of specified molecular fragments of the composite molecule. In the solid state trade, these are often called ''projections of the DOS'' or ''local DOS.'' Whatever they're called, they divide up the DOS among the atoms. The integral of these projections up to the Fermi level then gives the total electron density on a given atom or in a specific orbital. Then, by reference to some standard density, a charge can be assigned.

Figures 11 and 12 give the partitioning of the electron density between Pt and H in the PtH_4^{2-} stack, and between Ti and O in rutile. Everything is as **27** and **29** predict, as the chemist knows it should be; the lower orbitals are localized in the more electronegative ligands (H or O), the higher ones on the metal.

Do we want more specific information? In TiO_2 we might want to see the crystal field argument upheld. So we ask for the contributions of the

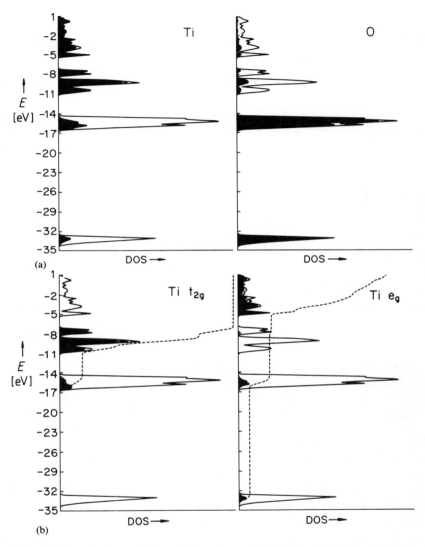

Figure 12 Contributions of Ti and O to the total DOS of rutile, TiO_2 are shown at top. At bottom, the t_{2g} and e_g Ti contributions are shown; their integration (on a scale of 0–100%) is given by the dashed line.

three orbitals that make up the t_{2g} (xz, yz, xy in a local coordinate system) and e_g (z^2, $x^2–y^2$) sets. This is also shown in Fig. 12. Note the very clear separation of the t_{2g} and e_g orbitals. The e_g has a small amount of density in the O 2s and 2p bands (σ bonding) and t_{2g} in the O 2p band (π bonding).

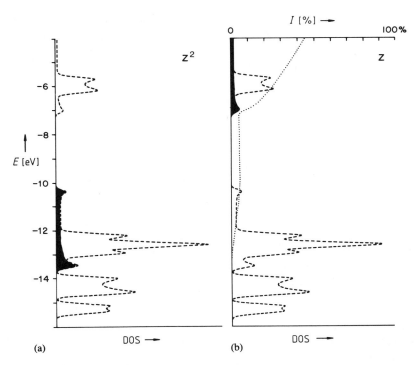

Figure 13 z^2 and z contributions to the total DOS of an eclipsed $PtH_4{}^{2-}$ stack. The dotted line is an integration of the z-orbital contribution.

Each metal orbital type (t_{2g} or e_g) is spread out into a band, but the memory of the near-octahedral local crystal field is very clear.

In $PtH_4{}^{2-}$ we could ask the computer to give us the z^2 contribution to the DOS, or the z part. If we look at the z component of the DOS in $PtH_4{}^{2-}$, we see a small contribution in the top of the z^2 band. This is most easily picked up by the integral in Fig. 13. The dotted line is a simple integration, like a nuclear magnetic resonance (NMR) integration. It counts, on a scale of 0–100%, what percentage of the specified orbital is filled at a given energy. At the Fermi level in unoxidized $PtH_4{}^{2-}$, 4% of the p_z states are filled.

How does this come about? There are two ways to talk about it. Locally, the donor function of one monomer (z^2) can interact with the acceptor function (z) of its neighbor. This is shown in **30**. The overlap is good, but the energy match is poor.[11] So the interaction is small, but it's there. Alternatively, one could think about interaction of the Bloch functions, or symmetry-adapted z and z^2 crystal orbitals. At $k = 0$ and π/a, they don't mix. But at every interior point in the Brillouin zone, the symmetry group of Ψ is isomorphic to C_{4v},[15], and both z and z^2 Bloch functions transform as a_1. So they mix. Some small bonding is provided by this mixing, but it is very small. When the stack is oxidized, the loss of this bonding (which would lengthen the Pt–Pt contact) is overcome by the loss of Pt–Pt antibonding that is a consequence of the vacated orbitals being at the top of the z^2 band.

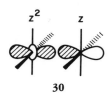

30

THE DETECTIVE WORK OF TRACING MOLECULE-SURFACE INTERACTIONS: DECOMPOSITION OF THE DOS

For another illustration of the utility of DOS decompositions, let's turn to a surface problem. We saw in a previous section the band structures and DOS of the CO overlayer and the Ni slab separately (Figs. 6, 7, 9). Now let's put them together in Fig. 14. The adsorption geometry is that shown earlier in 24, with Ni–C 1.8 Å. Only the densities of states are shown, based on the band structures of Figs. 7 and 9.[27] Some of the wriggles in the DOS curves also are not real, but a result of insufficient k-point sampling in the computation.

It's clear that the composite system $c(2 \times 2)$CO–Ni(100) is roughly a superposition of the slab and CO layers. Yet things have happened. Some of them are clear—the 5σ peak in the DOS has moved down. Some are less clear—where is the $2\pi^*$, and which orbitals on the metal are active in the interaction?

Let's see how the partitioning of the total DOS helps us to trace down the bonding in the chemisorbed CO system. Figure 15 shows the 5σ and $2\pi^*$ contributions to the DOS. The dotted line is a simple integration of the DOS of the fragment of contributing orbital. The relevant scale, 0–100%, is to be read at top. The integration shows the total percentage of the given orbital that's occupied at a specified energy. It is clear that the 5σ orbital, though pushed down in energy, remains quite localized. Its occupation (the integral of this DOS contribution up to the Fermi level) is 1.62 electrons. The $2\pi^*$ orbital obviously is much more delocalized. It is mixing with the metal d band and, as a result, there is a total of 0.74 electron in the $2\pi^*$ levels together.

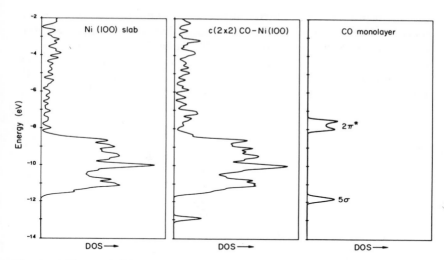

Figure 14 The total density of states of a model $c(2 \times 2)$CO–Ni(100) system (center), compared to its isolated four-layer Ni slab (left) and CO monolayer components.

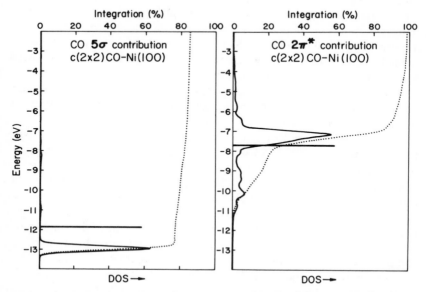

Figure 15 For the $c(2 \times 2)$CO–Ni(100) model this shows the 5σ and $2\pi^*$ contributions to the total DOS. Each contribution is magnified. The position of each level in isolated CO is marked by a line. The integration of the DOS contribution is given by the dotted line.

Which levels on the metal surface are responsible for these interactions? In discrete molecular systems we know that the important contributions to bonding are forward donation, **31a**, from the carbonyl lone pair 5σ to some appropriate hybrid on a partner metal fragment, and back donation, **31b**, involving the $2\pi^*$ of CO and a d_π orbital xz, yz of the metal. We would suspect that similar interactions are operative on the surface.

31

These can be looked for by setting side by side the $d_\sigma(z^2)$ and 5σ contributions to the DOS, and $d_\pi(xz, yz)$ and $2\pi^*$ contributions. In Fig. 16 the π interaction is clearest: note how $2\pi^*$ picks up density where the d_π states are, and vice versa, the d_π states have a "resonance" in the $2\pi^*$ density. I haven't shown the DOS of other metal levels, but were I to do so, it would be seen that such resonances are *not* found between those metal levels and 5σ and $2\pi^*$. The reader can confirm at least that 5σ does not pick up density where d_π states are, nor $2\pi^*$ where d_σ states are mainly found.[27] There is also some minor interaction of CO $2\pi^*$ with metal p_π states, a phenomenon not analyzed here.[28]

Let's consider another system in order to reinforce our comfort with these fragment analyses. In **25** we drew several acetylene–Pt(111) structures with coverage = 1/4. Consider one of these, the dibridged adsorption site alternative **25b** redrawn in **32**. The acetylene brings to the adsorption process a degenerate set of high-lying occupied π orbitals, and also an important unoccupied π^* set. These are shown at the top of **33**. In all known molecular and surface complexes, the acetylene is bent. This breaks the degeneracy of π and π^*, some s character mixing into the π_σ and π_σ^* components that lie in the bending plane and point to the surface. The valence orbitals are shown at the bottom of **33**. In Fig. 17 we show the contributions of these valence orbitals to the total DOS of **33**. The sticks mark the positions of the acetylene orbitals in the isolated molecule. It is clear that π and π^* interact less than π_σ and π_σ^* of CO.[29]

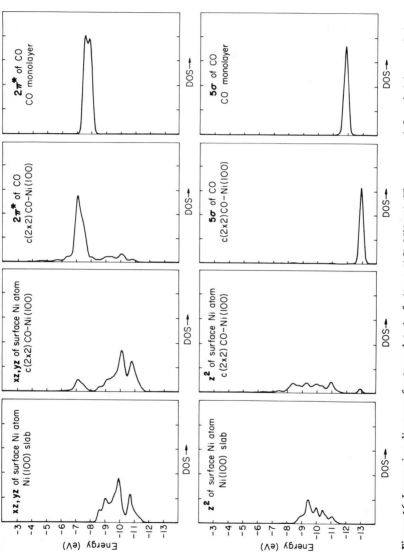

Figure 16 Interaction diagrams for 5σ and $2\pi^*$ of $c(2 \times 2)$C)–Ni(100). The extreme left and right panels in each case show the contributions of the appropriate orbitals (z^2 for 5σ, xz, yz for $2\pi^*$) of a surface metal atom (left) and of the corresponding isolated CO monolayer MO. The middle two panels then show the contributions of the same fragment MOs to the DOS of the composite chemisorption system.

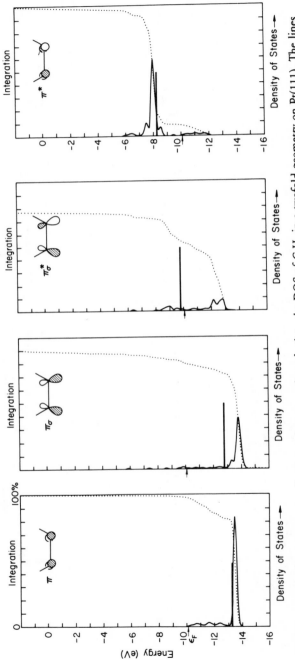

Figure 17 From left to right: contributions of π, π_σ, π_σ^*, and π^* to the DOS of C_2H_2 in a twofold geometry on Pt(111). The lines mark the positions of these levels in a free bent acetylene. The integrations of the DOS contributions are indicated by the dotted line.

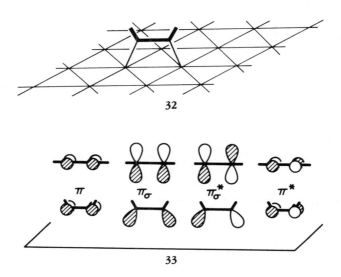

32

33

As for a third system: in the early stages of dissociative H_2 chemisorption, it is thought that H_2 approaches perpendicular to the surface, as in **34**. Consider Ni(111), related to the Pt(111) surface discussed earlier. Figure 18 shows a series of three snapshots of the total DOS and its $\sigma_u^*(H_2)$ projection.[30] These are computed at separations of 3.0, 2.5, and 2.0 Å from the nearest H of H_2 to the Ni atom directly below it. The σ_g orbital of H_2 (the lowest peak in the DOS in Fig. 18) remains quite localized. But the σ_u^* interacts and is strongly delocalized, with its main density pushed up. The primary mixing is with the Ni s, p band. As the H_2 approaches, some σ_u^* density comes below the Fermi level.

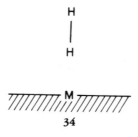

34

Why does σ_u^* interact more than σ_g? The classical perturbation theoretic measure of interaction:

$$\Delta E = \frac{|H_{ij}|^2}{E_i^0 - E_j^0}$$

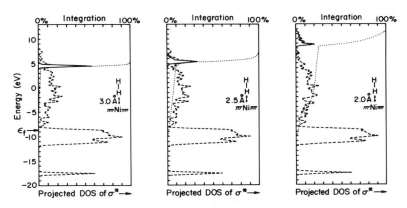

Figure 18 That part of the total DOS (dashed line) which is in the H_2 $\sigma_u{}^*$ (solid line) at various approach distances of a frozen H_2 to a Ni(111) surface model. The dotted line is an integration of the H_2 density.

helps one to understand this. $\sigma_u{}^*$ is more in resonance in energy, at least with the metal s, p band. In addition, its interaction with an appropriate symmetry metal orbital is greater than that of σ_g, at any given energy. This is the consequence of including overlap in the normalization:

$$\Psi_\pm = \frac{1}{\sqrt{2(1 \pm S_{12})}}\,(\phi_1 \pm \phi_2)$$

The $\sigma_u{}^*$ coefficients are substantially greater than those in σ_g. This has been pointed out by many individuals, but in the present context importantly emphasized by Shustorovich and Baetzold. [31-33]

We have seen that we can locate the electrons in the crystal. But...

WHERE ARE THE BONDS?

Local bonding considerations (see **27, 29**) trivially lead us to assign bonding characteristics to certain orbitals and, therefore, bands. There must be a way to find these bonds in the bands that a fully delocalized calculation gives.

It's possible to extend the idea of an overlap population to a crystal. Recall that in the integration of Ψ^2 for a two-center orbital, $2c_1 c_2 S_{12}$ was a characteristic of bonding. If the overlap integral is taken as positive (and it can always be arranged so), then this quantity scales as we expect of a bond order: it is positive (bonding) if c_1 and c_2 are of the same sign, and negative if c_1 and c_2 are of opposite sign. And the magnitude of the "Mulliken overlap population," for that is what $2c_1 c_2 S_{12}$ (summed over all orbitals on the two atoms, over all occupied MOs) is called, depends on c_i, c_j, S_{ij}.

Before we move into the solid, let's take a look at how these overlap populations might be used in a molecular problem. Figure 19 shows the familiar energy levels of a diatomic, N_2, a density-of-states plot of these (just sticks proportional to the number of levels, of length 1 for σ, 2 for π), and the contributions of these levels to the overlap population. $1\sigma_g$ and $1\sigma_u$ (not shown in the figure) contribute little because S_{ij} is small between tight 1s orbitals. $2\sigma_g$ is strongly bonding, $2\sigma_u$ and $3\sigma_g$ are essentially nonbonding. These are best characterized as lone pair combinations. π_u is bonding, π_g antibonding, $3\sigma_u$ the σ^* level. The right-hand side of Fig. 19 characterizes the bonding in N_2 at a glance. It tells us that maximal bonding is there for seven electron pairs (counting $1\sigma_g$ and $1\sigma_u$); more or fewer electrons will lower the N–N overlap population. It would be nice to have something like this for extended systems.

A bond indicator is easily constructed for the solid. An obvious procedure is to take all the states in a certain energy interval and interrogate them as to their bonding proclivities, measured by the Mulliken overlap population, $2c_i c_j S_{ij}$. What we are defining is an overlap population-weighted density of states. The beginning of the obvious acronym (OPWDOS) has unfortunately been preempted by another common usage in solid state physics. For that reason, we have called this quantity COOP, for crystal orbital overlap population.[34] It's also nice to think of the suggestion of orbitals working together to make bonds in the crystal, so the word is pronounced "co-op."

To get a feeling for this quantity, let's think about what a COOP curve for a hydrogen chain looks like. The simple band structure and DOS were given earlier, 26; they are repeated with the COOP curve in 35.

To calculate a COOP curve, one has to specify a bond. Let's take the nearest neighbor 1, 2 interaction. The bottom of the band is 1, 2 bonding, the middle nonbonding, the top antibonding. The COOP curve obviously has the shape shown at right in 35. But not all COOP curves look that way. If we specify the 1, 3 next nearest neighbor bond (silly for a linear chain, not so silly if the chain is kinked), then the bottom *and* the top of the band are 1, 3 bonding, the middle antibonding. That curve, the dashed line in the drawing 35, is different in shape. And, of course, its bonding and antibonding amplitude is much smaller because of the rapid decrease of S_{ij} with distance.

Note the general characteristics of COOP curves: positive regions that are bonding, negative regions that are antibonding. The amplitudes of these curves depend on the number of states in that energy interval, the

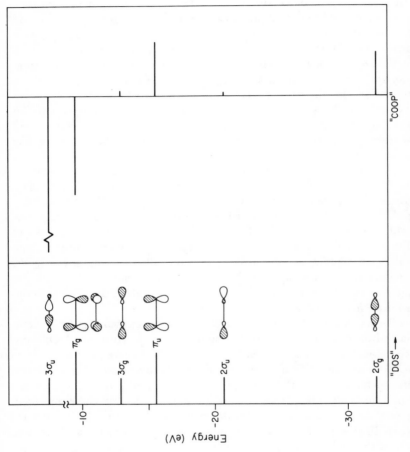

Figure 19 The orbitals of N_2 (left) and a "solid state way," to plot the DOS and COOP curves for this molecule. The $1\sigma_g$ and $1\sigma_u$ orbitals are out of the range of this figure.

magnitude of the coupling overlap, and the size of the coefficients in the MOs.

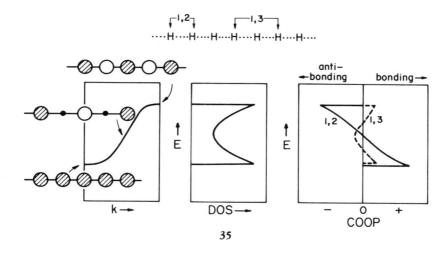

35

The integral of the COOP curve up to the Fermi level is the total overlap population of the specified bond. This points us to another way of thinking of the DOS and COOP curves. These are the differential versions of electronic occupation and bond order indices in the crystal. The integral of the DOS to the Fermi level gives the total number of electrons; the integral of the COOP curve gives the total overlap population, which is not identical to the bond order, but which scales like it. It is the closest a theoretician can get to that ill-defined but fantastically useful, simple concept of a bond order.

To move to something a little more complicated than the hydrogen or polyene chain, let's examine the COOP curves for the PtH_4^{2-} chain. Figure 20 shows both the Pt–H and Pt–Pt COOP curves. The DOS curve for the polymer is also drawn. The characterization of certain bands as bonding or antibonding is obvious, and matches fully the expectations of the approximate sketch **27**. The bands at -14, -15 eV are Pt–H σ bonding, the band at -6 eV Pt–H antibonding (this is the crystal field destabilized x^2-y^2 orbital). It is no surprise that the mass of d-block levels between -10 and -13 eV doesn't contribute anything to Pt–H bonding. But, of course, it is these orbitals that are involved in Pt–Pt bonding. The rather complex structure of the -10 to -13-eV region is easily understood by thinking of it as a superposition of σ (z^2-z^2), π (xz, yz)–(xz, yz), and δ (xy–xy) bonding and antibonding, as shown in **36**. Each type of bonding generates a band, the bottom of which is bonding and the top antibonding (see **35** and Fig. 3).

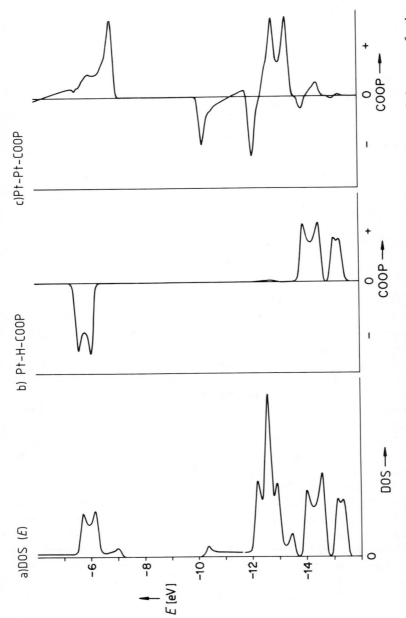

Figure 20 Total density of states (left), and Pt–H (middle) and Pt–Pt (right) crystal orbital overlap population curves for the eclipsed PtH_4^{2-} stack.

The δ contribution to the COOP is small because of the poor overlap involved. The large Pt–Pt bonding region at − 7 eV is due to the bottom of the Pt z band.

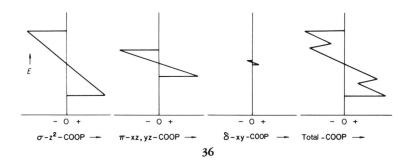

| σ-z²-COOP → | π-xz, yz-COOP → | δ-xy-COOP → | Total -COOP → |

36

We now have a clear representation of the Pt–H and Pt–Pt bonding properties as a function of energy. If we are presented with an oxidized material, then the consequences of the oxidation on the bonding are crystal clear from Fig. 20. Removing electrons from the top of the z² band at − 10 eV takes them from orbitals that are Pt–Pt antibonding, Pt–H nonbonding. So we expect the Pt–Pt separation, the stacking distance, to decrease as it does. [12]

The tuning of electron counts is one of the strategies of the solid state chemists. Elements can be substituted, atoms intercalated, nonstoichiometries enhanced. Oxidation and reduction, in solid state chemistry as in ordinary molecular solution chemistry, are about as characteristic (but experimentally not always trivial) chemical activities as one can conceive. The conclusions we reached for the Pt–Pt chain were simple, easily anticipated. Other cases are guaranteed to be more complicated. The COOP curves allow one, at a glance, to reach conclusions about the local effects on bond length (will bonds be weaker, stronger) upon oxidation or reduction.

Earlier we showed a band structure for rutile. The corresponding COOP curve for the Ti–O bond (Fig. 21) is extremely simple. Note the bonding in the lower oxygen bands and antibonding in the e_g crystal field destabilized orbitals. The t_{2g} band is, as expected, Ti–O nonbonding.

Let's try our hand at predicting the DOS for something quite different from PtH_4^{2-} or TiO_2, namely, a bulk transition metal, the face-centered-cubic Ni structure. Each metal atom has as its valence orbitals 3d, 4s, 4p, ordered in energy approximately as at the left in **37**. Each will spread out into a band. We can make some judgment as to the width of the bands from the overlap. The s, p orbitals are diffuse, their overlap will be large, and a

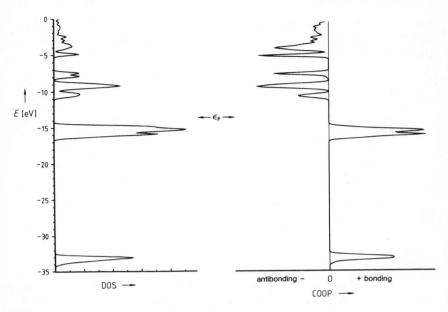

Figure 21 DOS and Ti–O COOP for rutile.

wide band will result. They also mix with each other extensively. The d orbitals are contracted, and so will give rise to a relatively narrow band.

The computed DOS for bulk Ni (bypassing the actual band structure) is shown in Fig. 22, along with the Ni s and p contributions to that DOS. What is not s or p is a d contribution. The general features of **37** are reproduced. At the Fermi level, a substantial part of the s band is occupied, so that the calculated[35] Ni configuration is $d^{9.15}s^{0.62}p^{0.23}$.

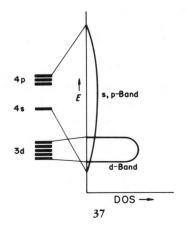

37

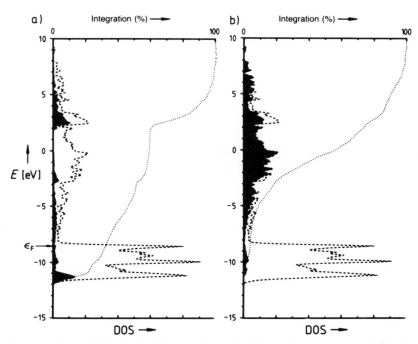

Figure 22 Total DOS (dashed line) and 4s and 4p contributions to it in bulk Ni. The dotted line is an integration of the occupation of a specified orbital, on a scale of 0–100% given at top.

What would one expect of the COOP curve for bulk Ni? As a first approximation, we could generate the COOP curve for each band separately, as in **38a** and **b**. Each band in **37** has a lower Ni–Ni bonding part, an upper Ni–Ni antibonding part. The composite is **38c**. The computed COOP curve is in Fig. 23. The expectations of **38c** are met reasonably well.

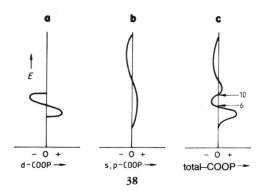

38

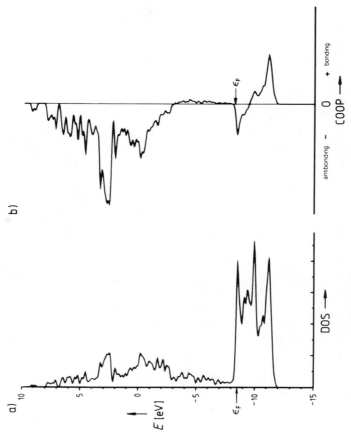

Figure 23 The total DOS and nearest neighbor Ni–Ni COOP in bulk Ni.

A metal–metal COOP curve like that of **38c** or Fig. 23 is expected for any transition metal. The energy levels might be shifted up, they might be shifted down, but their bonding characteristics are likely to be the same. If we assume that a similar band structure and COOP curve hold for all metals (in the solid state trade this would be called the rigid band model), then Fig. 23 gains tremendous power. It summarizes, simply, the cohesive energies of all metals. As one moves across the transition series, the M–M overlap population (which is clearly related to the binding or cohesive energy) will increase, peaking at about 6 electrons/per metal—Cr, Mo, W. Then it will decrease toward the end of the transition series and rise again for small s, p electron counts. For more than 14 electrons, a metal is unlikely; the net overlap population for such high coordination becomes negative. Molecular allotropes with lower coordination are favored. There is much more to cohesive energies and the metal–nonmetal transition than this. Still, a lot of physics and chemistry flows from the simple construction of **38**.

COOP curves are a useful tool in the tracing down of surface-adsorbate interactions. Let's see, for instance, how this indicator may be used to support the picture of CO chemisorption that was described above. The relevant curve is in Fig. 24. The solid line describes Ni–C bonding, the dotted line C–O bonding. The C–O bonding is largely concentrated in orbitals that are out of the range of (below) this figure. Note the major contribution to Ni–C bonding in both the 5σ peak and the bottom of the d band. The 5σ contribution is due to σ-bonding, **31a**. But the bottom of the d band contributes through π-bonding, **31b**. This is evident from the "mirroring" C–O antibonding in the same region. The antibonding component of that $d_\pi-2\pi^*$ interaction is responsible for the Ni–C and C–O antibonding above the Fermi level. [27]

It may be useful to emphasize that these curves are not only descriptive, but also form a part of the story of tracing down interaction. For instance, supposing we were not so sure that it is the $d_\pi-2\pi^*$ interaction that is responsible for a good part of the bonding. Instead, we could have imagined π bonding between 1π and some unfilled d_π orbitals. The interaction is indicated schematically in **39**. If this mixing were important, the d-block orbitals, interacting in an antibonding way with 1π below them, should become in part Ni–C *antibonding* and C–O *bonding*. Nothing of this sort is seen in Fig. 24. The C–O antibonding in the d-block region is, instead, diagnostic of $2\pi^*$ mixing being important.

Incidentally, the integrated overlap populations up to the Fermi level are Ni–C 0.84, C–O 1.04. In free CO the corresponding overlap population is 1.21. The bond weakening is largely due to population of $2\pi^*$ on chemisorption.

Another illustration of the utility of COOP curves is provided next by a question of chemisorption site preference. On many surfaces, including

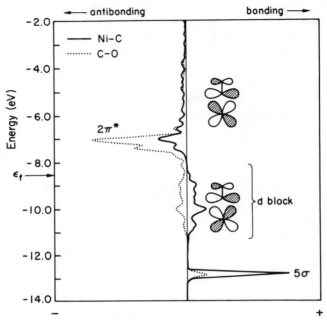

Figure 24 Crystal orbital overlap population for CO, on top, in a $c(2 \times 2)$CO–Ni(100) model. Representative orbital combinations are drawn out.

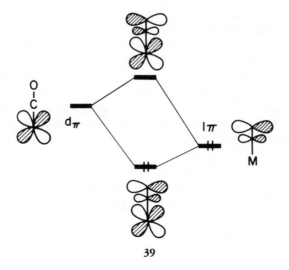

39

Pt(111), a particularly stable dead end in the surface chemistry of acetylene is ethylidyne (CH$_3$).[36] How that extra hydrogen is picked up is a fascinating question. But let's bypass that and think about where the CCH$_3$ wants to be. Diagram **40** shows three alternatives: one-fold or "on top," two-fold or "bridging," and three-fold or "capping." Experiment and theory show a great preference for the capping site. Why?

40

41

The important frontier orbitals of a carbyne, CR, are shown in **41**. The C 2p orbitals, the e set, are a particularly attractive acceptor set, certain to be important in any chemistry of this fragment. We could trace its involvement in the three alternative geometries **40** via DOS plots, but instead we choose to show in Fig. 25 the Pt–C COOP curve for one-fold and three-fold adsorption.

In both on-top and capping sites the carbyne e set finds metal orbitals with which to interact. Bonding and antibonding combinations form. The coupling overlaps are much better in the capping site. The result is that the carbon–metal e-type antibonding combinations do not rise above the Fermi level in the one-fold case but do so in the three-fold case. Figure 25 clearly shows this—the bonding and antibonding combinations are responsible for recognizable positive and negative COOP peaks. The total surface–CCH$_3$ overlap populations are 0.78 in the one-fold case, 1.60 in the three-fold case. The total energy follows these bonding considerations; the capping site is much preferred.[29]

With a little effort, we have constructed the tools—density of states, its decompositions, the crystal orbital overlap population—that allow us to move from a complicated, completely delocalized set of crystal orbitals or Bloch functions to the localized, chemical description. There is no mystery

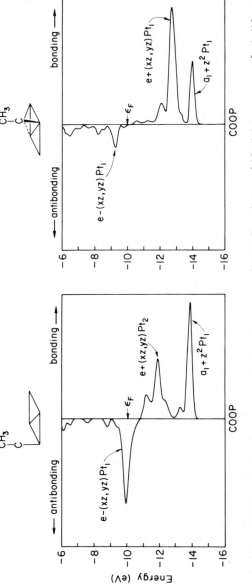

Figure 25 COOP curve for the α–carbon–Pt$_1$ bond in the one–fold (left) and three–fold (right) geometry of ethylidyne, CCH$_3$, on Pt(111).

in this motion. In fact, what I hope I have shown here is just how much power there is in the chemists' concepts. The construction of the *approximate* DOS and bonding characteristics of a PtH$_4{}^{2-}$ polymer, or rutile, or bulk Ni, is really easy.

Of course, there is much more to solid state physics than band structures. The mechanism of conductivity, the remarkable phenomenon of superconductivity, the multitude of electric and magnetic phenomena that are special to the solid state, for these one needs the tools and ingenuity of physics.[9] But as for *bonding* in the solid state, I think (some will disagree) there is nothing new, only a different language.

A SOLID STATE SAMPLE PROBLEM: THE ThCr$_2$Si$_2$ STRUCTURE

The preceding sections have outlined some of the theoretical tools for analysis of bonding in the solid state. To see how these ideas can be integrated, let's discuss a specific problem.

More than 200 compounds of AB$_2$X$_2$ stoichiometry adopt the ThCr$_2$Si$_2$ type structure.[37] However, you are not likely to find any mention of these in any modern textbook of general inorganic chemistry, which tells us something about the ascendancy of molecular inorganic chemistry, especially transition metal organometallic chemistry, in the last three decades. However, these compounds are there, we know their structures, and they have interesting properties. A is typically a rare earth, alkaline earth, or alkali element, B is a transition metal or main group element, and X comes from groups 15, 14, and occasionally 13. Since the synthesis of AB$_2$X$_2$ with A = a rare earth element, by Parthé, Rossi, and their coworkers, the unusual physical properties exhibited by these solids have attracted much attention. Physicists speak with enthusiasm of valence fluctuation, p-wave or heavy fermion superconductivity, and of many peculiar magnetic properties of these materials. The very structure of these materials carries much that is of interest to the chemist.

The ThCr$_2$Si$_2$ structure type for AB$_2$X$_2$ stoichiometry compounds is shown in **42**. It consists of B$_2$X$_2{}^{2-}$ layers interspersed with A^{2+} layers. The bonding between A and B$_2$X$_2$ layers appears largely ionic, which is why we write the charge partitioning as A^{2+} and B$_2$X$_2{}^{2-}$. But in the B$_2$X$_2{}^{2-}$ layer there is indication not only of covalent B–X bonding, but also some metal–metal B–B bonding. Typical metal–metal distances are in the range of 2.7–2.9 Å.

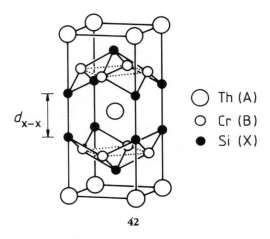

d_{x-x}

◯ Th (A)
○ Cr (B)
● Si (X)

42

A way to describe the B_2X_2 layers in these compounds is to imagine a perfect square planar two-dimensional lattice of metal atoms, above and below the fourfold hollows of which lie the main group X atoms. This is shown in **43**. The coordination environment of the metal (B) is approximately tetrahedral in the main group elements (X), with four additional square planar near-neighbor metals. The coordination of the X atoms is much more unusual; they reside at the apex of a square pyramid.

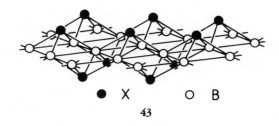

● X   ○ B

43

It may be noted here that there are alternative ways to describe the layer structure. For instance, the B_2X_2 layer may be thought of as being built up by sharing four of the six edges of a BX_4 tetrahedron by infinite extension in two dimensions, as in **44**. Such packing diagrams or alternative ways of looking at the same structure are inherently useful; a new view often leads to new insight. I would just introduce a very personal prejudice, voiced in the introduction, for views of structure that make as many connections as possible to other subfields of chemistry. On that basis, I would give a slight preference to **43** over **44**—the latter pulls one a little away from bonds.

There is a long X···X contact within the layer, but what becomes the main focus of this section is a remarkable tunable $X \cdots X$ contact *between* all layers, along the edges (and across top and bottom faces) of the tetragonal unit cell **42**. This contact (d_{x-x}) is the primary geometric variable in these structures.

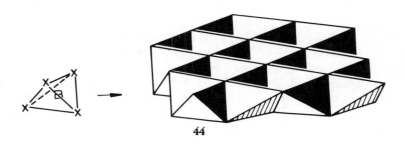

44

Sometimes d_{x-x} is long, sometimes it is short. In Table 1 are shown two series of compounds studied by Mewis.[38] In these the cation is kept constant, and so is the main group element P. Only the metal varies.

For reference the P–P distance is 2.21 Å in P4 and 2.192 Å in Me₂P–PMe₂. The P–P single-bond distance in many compounds is remarkably constant at 2.19–2.26 Å. The P=P double bond and P≡P triple bond lengths are around 2.03 and 1.87 Å respectively. It is clear that the short distances in the ThCr₂Si₂-type phosphides are characteristic of a full P–P single bond. The long contacts, such as 3.43 Å, imply essentially no bonding at all. All the compounds known with a nonbonding $X \cdots X$ separation contain metals from the left-hand side of the Periodic Table. In fact, examination of all the structures reveals a trend. As one moves left to right in the transition series, the P–P contact shortens. Clearly there is an electronic effect of work here; a $P \cdots P$ bond is made or broken in the solid state. We would like to understand how and why this happens.

Incidentally, let's see what happens if one takes a Zintl viewpoint of these structures. The long $P \cdots P$ contact would be associated with a filled octet P^{3-}, the full P–P single bond with a $P-P^{4-}$. For a divalent A^{2+} we

Table 1 X–X Distance in Some Phosphide Compounds of AB_2X_2 Type

AB_2X_2	d_{X-X} (Å)	AB_2X_2	d_{X-X} (Å)
$CaCu_{1.75}P_2$	2.25	$SrCu_{1.75}P_2$	2.30
$CaNi_2P_2$	2.30	$SrCo_2P_2$	3.42
$CaCo_2P_2$	2.45	$SrFe_2P_2$	3.43
$CaFe_2P_2$	2.71		

would be left with a metal in oxidation state II for the case of no P···P bond, oxidation state I for a single P–P bond. One could make some sense of the trend in terms of the energetics of the various metal oxidation states, but one way or another the Zintl picture has a difficult time with intermediate distances. How does one describe a P···P bond length of 2.72 Å? A delocalized approach has no problems with describing such partial bonding.

Chong Zheng and I[39] approached the AB$_2$X$_2$ structure, represented by a typical BaMn$_2$P$_2$ compound, in stages. First we looked at a single two-dimensional Mn$_2$P$_2$$^{2-}$ layer. Then we formed a three-dimensional Mn$_2$P$_2$$^{2-}$ sublattice by bringing many such layers together in the third dimension.

Consider a single Mn$_2$P$_2$ layer, **43**. The Mn–P distance is 2.455 Å, and the Mn–Md distance in the square metal lattice is 2.855 Å. The latter is definitely in the metal–metal bonding range, so a wide-band, delocalized picture is inevitable. But in some hierarchy or ranking of interactions, it is clear that Mn–P bonding is stronger than Mn–Mn. So let's construct this solid conceptually or think of it in terms of first turning on Mn–P bonding, and then the Mn–Mn interaction.

The local coordination environment at each Mn is approximately tetrahedral. If we had a discrete tetrahedral Mn complex, e.g., Mn(PR$_3$)$_4$, we might expect a qualitative bonding picture such as **45**. Four phosphine lone pairs, a$_1$ + t$_2$ in symmetry, interact with their symmetry match, mainly Mn 4s and 4p, but also with the t$_2$ component of the Mn 3d set. Four orbitals, mainly on P, P–Mn σ bonding, go down. Four orbitals, mainly on Mn, P–Mn σ antibonding, go up. The Mn d block splits in the expected two below three way.

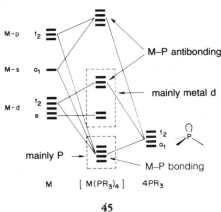

45

Something like this *must* happen in the solid. In addition, there are Mn–Mn bonding contacts in the layer, and these will lead to dispersion in those bands that are built up from orbitals containing substantial metal character. The combined construction is shown in Fig. 26.

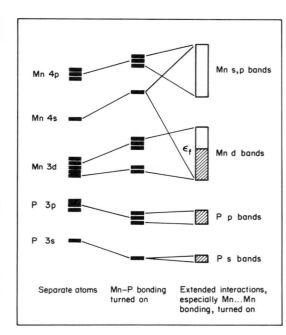

Figure 26 A schematic picture of the $Mn_2P_2{}^{2-}$ layer band structure as derived by first turning on local Mn–P interactions and then the two-dimensional periodicity and Mn–Mn interactions. The unit cell contains two Mn and two P atoms, so in reality each of the levels in the first two columns should be doubled.

Can we see this local, very chemical bonding construction in a delocalized band structure? Most certainly. The calculated (extended Hückel) band structure and total density of states of a single $Mn_2P_2{}^{2-}$ layer is illustrated in Fig. 27.

The unit cell is a rhomboid of two Mn and two P atoms. P is clearly more electronegative than Mn, so we expect two mainly P 3s bands below six P 3p bands below 10 Mn 3d bands. The number of bands in Fig. 27 checks. A decomposition of the DOS (Fig. 28) confirms the assignment.

What about the bonding characteristics predicted by the qualitative bonding scheme **45**? This is where a COOP curve is useful, as shown in Fig. 29. Note that the two lower bands (at -15 and -19 eV), which by the previous decomposition were seen to be mainly P, are Mn–P bonding, whereas the mainly metal bands around -12 eV are Mn–P nonbonding. The bunch of levels at approximately -9 eV is Mn–P antibonding—it corresponds to the crystal field destabilized t_2 level in **45**. The bottom of the mainly metal band is Mn–Mn bonding, the top Mn–Mn antibonding. Everything is as expected.

An interesting, slightly different approach to the bonding in the layer is obtained if we, so to speak, turn on Mn–Mn bonding first, then turn on Mn–P bonding by "inserting" or "intercalating" a P sublattice. This is

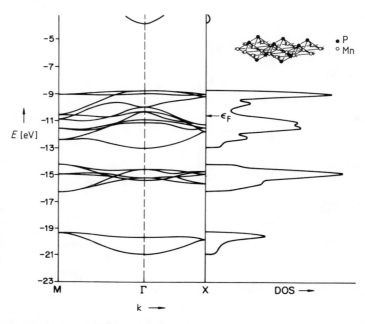

Figure 27 Band structure and DOS of a single Mn₂P₂²⁻ layer.

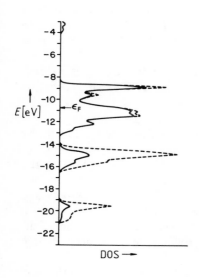

Figure 28 Total DOS of the composite Mn₂P₂²⁻ layer lattice (dashed line) and the contributions of Mn orbitals to that DOS (solid line). What is not on Mn is on P.

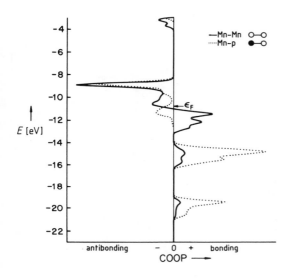

Figure 29 Crystal orbital overlap population curves for the Mn–Mn bonds (solid line) and Mn–P bonds (dotted line) in the $Mn_2P_2{}^{2-}$ single layer.

done in Fig. 30. At left is the P sublattice. We see P 3s (around − 19 eV) and P 3p (around − 14 eV) bands. Both are narrow because the P atoms are ∼ 4 Å apart. The Mn sublattice (middle of Fig. 30) shows a nicely dispersed density of states (DOS). The Mn–Mn separation is only 2.855 Å. Thus we have a two-dimensional metal, with a familiar wide s, p plus narrow d band

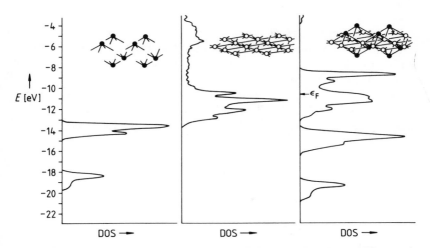

Figure 30 Total DOS of the P sublattice (left), the Mn sublattice (middle), and the composite $Mn_2P_2{}^{2-}$ layer lattice (right).

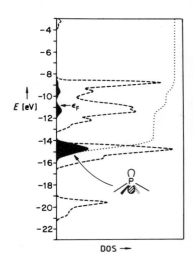

Figure 31 Phosphorus 3p$_z$ orbital contribution (dark area) to the total DOS (dashed line) of the Mn$_2$P$_2^{2-}$ single layer. The dotted line is an integration of the dark line, on a scale of 0–100%.

pattern. The bottom part of the DOS in the middle of Fig. 30 is the 3d band, the top is the lower part of the 4s, 4p band. At the right in Fig. 30 is the density of states of the composite Mn$_2$P$_2$ layer. Note how the individual P and Mn bunches of states repel each other on forming the composite lattice. Note also how part of the Mn d band stays where it is and part moves up. Here is the memory, within this delocalized structure, of the local e below t$_2$ crystal field splitting. There is no more graphic way of showing that what happens in the inorganic solid is similar to what happens in an isolated inorganic molecule.

Here is another, more chemical detail. Each phosphorus in the slab is in an unusual coordination environment, at the apex of a square pyramid of Mn atoms. A chemist looks for a lone pair, **46**, pointing away from the ligands. We can look for it, theoretically, by focusing on its directionality. P 3p$_z$ should contribute most to this lone pair, so we interrogate the DOS for its z contribution (Fig. 31). The p$_z$ orbital is indeed well localized, 70% of it in a band at approximately -15 eV. Here we see the lone pair.

46

A point that can be made here is that localization in energy space (such as we see for the P p$_z$ projection) implies localization in real space. The

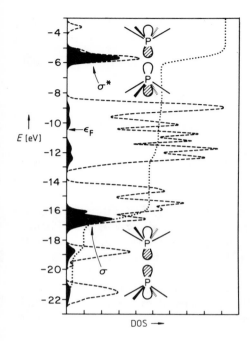

Figure 32 Phosphorus 3p$_z$ orbital contribution (dark area) to the total DOS (dashed line) of the three-dimensional Mn$_2$P$_2$$^{2-}$ lattice. The phosphorus-phosphorus bond length here is 2.4 Å. The dotted line is an integration, on a scale of 0–100%, of the 3p$_z$ orbital occupation.

easiest way to think this through is to go back to the construction of bands at the beginning of this book. The molecular orbitals of a crystal are always completely delocalized Bloch functions. But there is a difference between what we might call symmetry-enforced delocalization (formation of Bloch functions, little overlap) and real, chemical delocalization (overlap between unit cells). The former gives rise to narrow bands, the latter to highly dispersed ones. Turning the argument around, the existence of narrow bands is a sign of chemical localization, whereas wide bands imply real delocalization.

On to the three-dimensional solid. When the two-dimensional Mn$_2$P$_2$$^{2-}$ layers are brought together to form the three-dimensional solid (Mn$_2$P$_2$$^{2-}$, still without the counterions), the P 3p$_z$ orbitals or lone pairs in one layer form bonding and antibonding combinations with the corresponding orbitals in the layers above or below. Figure 32 shows the P 3p$_z$ density of states at interlayer P–P = 2.4 Å. The wide band at -8 to -12 eV is Mn 3d. Below and above this metal band are P bands, and in these, quite well localized, are P–P σ and σ* combinations, **47**. These bands are narrow because the lateral P–P distance is long.

Perhaps it's appropriate to stop here and reflect on what has happened. There are Avogadro's number of levels per atomic orbital in the solid. It's all delocalized bonding, but with our theoretical tools we have been able to see, quite localized in energy, orbitals of a diatomic molecule. The localization in energy reflects the validity of a localization in space, i.e., a bond.

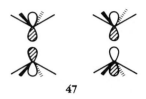

47

If the three-dimensional calculation is repeated at different interslab or P$\cdots$P distances, all that happens is that the localized P–P and σ^* bands occur at different energies. Their splitting decreases with increasing P$\cdots$P separation, as one would expect from their respective bonding and antibonding nature.

We are now in a position to explain simply the effect of the transition metal on the P–P separation. What happens when the transition metal moves to the right-hand side of the Periodic Table? The increased nuclear charge will be more incompletely screened and the d electrons more tightly bound. As a result, the d band comes down in energy and becomes narrower. At the same time, the band filling increases as one moves to the right in the transition series. The balance is complicated, and it is important. Diagram **48** shows the result. For details the reader is referred to the definitive work of O. K. Andersen.[40]

Diagram **48** is the most important single graph of metal physics. It is analogous in its significance to the plot of the ionization potentials of atoms or diatomic molecules. At the right side of the transition series, which is our area of concern, the Fermi level falls as one moves to the right, and the work function of the metal increases.

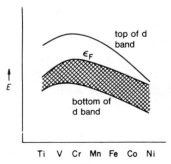

48

Now imagine superimposed on this variable-energy sea of electrons the P–P and σ^* bands for some typical, moderately bonding P–P distance, **49**. In the middle of the transition series, the metal Fermi level is above the P–P σ^*. Both σ and σ^* are occupied, so there is no resultant P–P bond. As P–P stretches in response, the σ^* only becomes more filled. On the right side of the transition series, the P–P σ^* is above the Fermi level of the metal, and so is unfilled. The filled P–P σ makes a P–P bond. Making the P–P distance shorter only improves this situation.

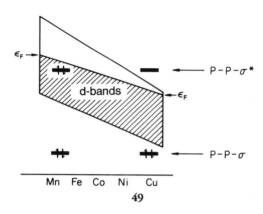

49

The steady, gradual variation of the P–P distance would seem to be as inconsistent with the molecular orbital model shown here as it was with the Zintl concept. This is not so. If we turn on the interaction between the P atoms and the metal layer (and we have seen before that this interaction is substantial), we will get a mixing of P and Mn orbitals. The discontinuity of the above picture (either single-bond or no bond) will be replaced by a continuous variation of P σ and σ^* orbitals' occupation between 2 and 0.

The experimentally observed trend has been explained. There is much more to the AB_2X_2 structures than I have been able to present here,[37,39] of course. More important than the rationalizations and predictions of the experimental facts that one is able to make in this case is the degree of understanding one can achieve and the facility of motion between chemical and physical perspectives.

THE FRONTIER ORBITAL PERSPECTIVE

The analytical tools for moving *backward* from a band calculation to the underlying fundamental interactions are at hand. Now let's discuss the motion in the *forward* direction, the model of orbitals and their interaction, as analyzed by perturbation theory. In a sense we already used this in **27, 29** and Fig. 26, i.e., the mental construction of what we anticipated in building the Mn_2P_2 layer.

This is the frontier orbital picture.[11,41] A chemical interaction (between two parts of a molecule) or reaction (between two molecules) can be analyzed from the starting point of the energy levels of the interacting fragments or molecules. The theoretical tool one uses is perturbation theory. To second order, the interactions between two systems are pairwise additive over the MOs and each pair interaction is governed by the expression:

$$\Delta E = \frac{|H_{ij}|^2}{E_i° - E_j°}$$

That's what a squiggly line in the interaction diagram **50** indicates.

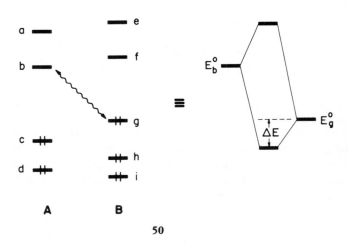

50

Individual interactions may be classified according to the total number of electrons in the two orbitals involved; thus ① and ② in **51** are two-electron, ③ is four-electron, ④ is zero-electron. ① and ② are clearly stabilizing (see the right side of **50**). This is where true bonding is found, with its range between covalent (orbitals balanced in energy and extent in space) or dative (orbitals unequal partners in interaction, charge transfer from donor to acceptor an inevitable correlate of bonding). Interaction ④ has no direct energetic consequences, since the bonding combination is unoccupied. And interaction ③ is repulsive because what happens when the overlap is included in the calculations (**52**) is that the antibonding combination goes up more than the bonding one goes down. The total energy is greater than that of the separate isolated levels. [11]

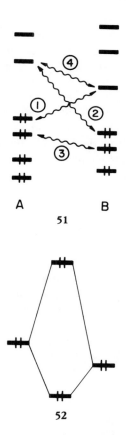

The electronic energy levels of molecules are separated by energies of the order of an electron volt. This makes them quantum systems *par excellence* and allows the singling out of certain levels as controlling a geometric preference or a reactivity. For instance, in **50** acceptor level $|b\rangle$ of fragment A is closer in energy to donor level $|g\rangle$ of fragment B, compared to $|h\rangle$ and $|i\rangle$. If it should happen that the overlaps $\langle b|h\rangle$ and $\langle b|i\rangle$ are also much smaller than $\langle b|g\rangle$, then both the numerator and the denominator of the perturbation expression single out the $b(A)-g(B)$ interaction as an important, perhaps the most important, one. In general, it turns out that the highest occupied molecular orbital (HOMO) or a small subset of higher lying levels, and the lowest unoccupied molecular orbital (LUMO) or some subset of unoccupied MO's, dominate the interaction between two molecules. These are called the frontier orbitals. They are the valence orbitals of the molecule, the orbitals most easily perturbed in any molecular interaction. In them resides control of the chemistry of the molecule.

It should be realized that this description, while of immense interpretive power, is only a one-electron model. To analyze orbital interactions properly, in a many-electron way, is not easy. The simple picture of **51** seems to be lost; competing interaction or partition schemes have been suggested.[42] One way to appreciate the problem a true many-electron theory has in analyzing interactions is to realize that the energy levels of A and B are not invariant to electron transfer. They change in energy depending on the charge: on fragments A and B a positive charge makes all the energy levels go down, and a negative charge makes them go up. Realizing this, one has learned the most important correction to the simple one-electron picture.

ORBITAL INTERACTION ON A SURFACE

It is now clear that the apparatus of densities of states and crystal orbital overlap populations has served to restore to us a frontier orbital or interaction diagram way of thinking about the way molecules bond to surfaces, or the way atoms or clusters bond in three-dimensional extended structures. Whether it is $2\pi^*$ CO with d_π of Ni(100), or e of CR with some part of the Pt(111) band, or the Mn and P sublattices in $Mn_2P_2^{2-}$, or the Chevrel phases discussed below, in all of these cases we can describe what happens in terms of local action. The only novel feature so far is that the interacting orbitals in the solid often are not single orbitals localized in energy or space, but bands.

A side-by-side comparison of orbital interactions in discrete molecules and a molecule with a surface is revealing. Diagram **53** is a typical molecular interaction diagram, **54** a molecule-surface one. Even though a molecule is, in general, a many-level system, let's assume, in the spirit of a frontier orbital analysis, that a small set of frontier orbitals dominates. This is why the squiggly lines symbolizing interaction go to the HOMO and LUMO of each component.

Within a one-electron picture, the following statements can be made (and they apply to both the molecule and the surface unless specifically indicated not to do so).

(1) The controlling interactions are likely to be the two-orbital, two-electron stabilizing interactions ① and ②. Depending on the relative

energy of the orbitals and the quality of the overlap, each of these interactions will involve charge transfer from one system to the other. In interaction ①, A is the donor or base, and B, or the surface, is the acceptor or acid. In interaction ② these roles are reversed.

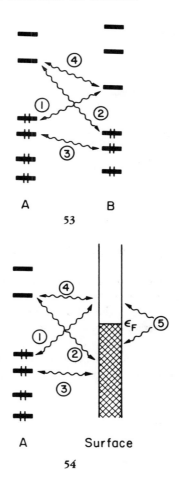

A   B

53

A   Surface

54

(2) Interaction ③ is a two-orbital, four-electron one. It is destabilizing, repulsive, as **55a** shows. In one-electron theories, this is where steric effects, lone pair repulsions, and the like are to be found.[11,41] These interactions may be important. They may prevent bonding interactions ① and ② from being realized. There is a special variant of this interaction that may occur in the solid but is unlikely to occur in discrete molecules. This is sketched in **55b**—the antibonding component of a four-electron, two-

orbital interaction may rise above the Fermi level. It will dump its electrons at the Fermi level and can no longer destabilize the system. Only the intersystem bonding combination remains filled.

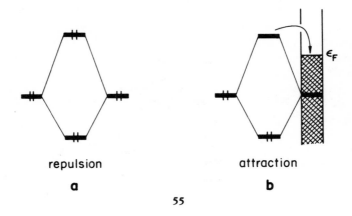

repulsion   attraction

a   b

55

The effect on molecule-surface bonding is clear—it is improved by this situation. What happens in the surface is less clear; let's defer discussion until we get to interaction ⑤.

(3) Interaction ④ involves two empty orbitals. In general, it would be discounted as having no energetic consequences. This is strictly true in molecular cases, **56a**. But in the solid, where there is a continuum of levels, the result of such interaction may be that the bonding combination of the two interacting levels falls below the Fermi level, **56b**. Becoming occupied, it will enhance fragment A–surface bonding. Again, there may be an effect on the surface because it has to supply the electrons for the occupation of that level.

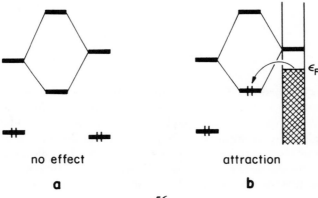

no effect   attraction

a   b

56

(4) Interaction ⑤ is something special to the metallic solid that comes from the states of the metal surface forming a near continuum. The interaction describes the second order energetic and bonding consequences of shifts of electron density around the Fermi level. First order interactions ①, ②, ③, and ④ will all move metal levels up and down. These metal levels, the ones that move, will belong to the atoms on the surface interacting with the adsorbate. The Fermi level remains constant—the bulk and surface are a nice reservoir of electrons. Therefore electrons (holes) will flow in the surface and in the bulk underneath it in order to compensate for the primary interactions. However, these compensating electrons or holes are not innocent of bonding themselves. Depending on the electron filling, they may be bonding or antibonding in the bulk, between surface atoms not involved with the adsorbate, even in surface atoms so involved, but in orbitals that are not used in bonding to the chemisorbed molecule.

Before leaving this section, I should like to say quite explicitly that there is little that is novel in the use my coworkers and I have made of interaction diagrams and perturbation theory applied to surfaces. A. B. Anderson[43] has consistently couched his explanations in that language, and so have Shustorovich and Baetzold[31,32]; Shustorovich's account of chemisorption is based on an explicit perturbation theoretical model.[44a] There is a very nice, quite chemical treatment of such a model in the work of Gadzuk,[44b] based on earlier considerations by Grimley.[45] van Santen[46a] draws interaction diagrams quite analogous to ours. Lowe has recently discussed frontier crystal orbitals explicitly[46b], and the interesting concept of interaction orbitals has recently been extended to surfaces by Fujimoto[46c]. Salem and his coworkers[47] developed a related perturbation theory based on a way of thinking about catalysis that includes a discussion of model finite Hückel crystals, privileged orbitals, generalized interactions diagrams, and the dissolution of adsorbate into catalyst bands. Other workers have also discussed interaction diagrams, privileged orbital sets, or orbital symmetry considerations in the solid.[48]

Let's make these interactions and interactions diagrams come to life through some specific applications.

A CASE STUDY: CO ON Ni(100)

The Ni(100)–CO system already discussed[26] seemed to provide an excellent example of the primary two-electron interactions at work. We found charge transfer from 5σ (its population going from 2.0 in the free CO to 1.62 in the CO-surface complex) and back donation from the surface to $2\pi^*$ (whose population rose from 0 to 0.74). Actually, there is an interesting wrinkle here in that the four- and zero-electron interactions mentioned in point 3 above manifest themselves.

To set a basis for what we will discuss, let's prepare a model molecular system for comparison. We'll build a metal–carbonyl bond between a d^6 ML_5 system and a carbon monoxide. The interaction diagram, **57**, will be familiar to a chemist; the acceptor function of the ML_n fragment is provided by a low-lying dsp hybrid.[11,49] The two-electron bonding interactions are quite explicit. They result (M = Ni, L = H$^-$, M–H 1.7 Å, M–CO 1.9 Å) in a depopulation of 5σ by 0.41 and a population of $2\pi^*$ by 0.51 electron. The metal functions involved in these interactions react correspondingly: so xz, yz loses 0.48 electron, and the hybrid orbital gains 0.48. The net charge drifts are pretty well described by the sum of what happens in these orbitals: CO as a whole gains 0.01e$^-$, and the ML_n fragment loses the same. The information is summarized in Table 2.

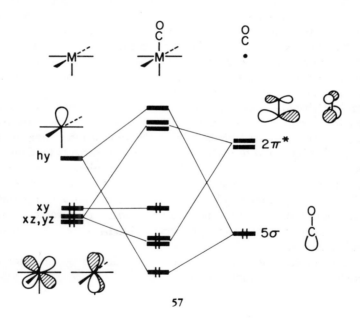

57

If one just looks at the CO, what happens on the surface seems to be similar, as noted above. And the d_π orbitals xz, yz are depopulated in $c(2 \times 2)$CO–Ni(100). But the d_σ, the z^2, the surface analogue of the hybrid, actually *loses* electron density on chemisorption of CO.

What is happening here is that the CO 5σ is interacting with the entire z^2 band, but perhaps more with its bottom, where the coupling overlap is greater. The z^2 band is nearly filled (1.93e in the metal slab). The net 5σ–d_σ band interaction would be repulsive, mainly due to four-electron two-

Table 2 Some Electron Densities in a Model H_3NiCO^- and the $c(2 \times 2)CO$–Ni(100) System

	NiH_5^-	$NiH_5(CO)^-$	CO		Ni(100)	$c(2 \times 2)CO$–Ni(100)	CO
5σ	—	1.59	2.0	5σ	—	1.62	2.0
$2\pi^*$	—	0.51	0.0	$2\pi^*$	—	0.74	0.0
hy	0.0	0.48	—	$d_\sigma{}^a$	1.93	1.43	—
d_π	4.0	3.52	—	$d_\pi{}^a$	3.81	3.31	—
CO	—	10.01	10.0	CO	—	10.25	10.0
H_5Ni	16.0	15.99	—	Ni^a	10.17^b	9.37	—

[a] For surface atoms that have CO on them.

[b] This number is not 10.0 because the surface layer of the slab is negative relative to the inner layer.

orbital interactions, were it not for the pushing of some antibonding combinations above the Fermi level (see **58** for a schematic). The net result is some loss of z^2 density and concomitant bonding.[50]

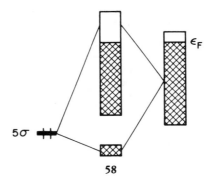

58

Where do those "lost" electrons go? Table 2 indicates that some, but certainly not all, go to the CO. Many are "dumped" at the Fermi level into orbitals that are mainly d band, but on the inner metal atoms, or on surface atoms not under CO. We will return to the bonding consequences of these electron interaction ⑤, in a while.

Before leaving this instructive example, I trust that the point is not lost that the primary bonding interactions ① and ② are remarkably alike in the molecule and on the surface. These forward and back donations are, of course, the consequence of the classical Dewar–Chatt–Duncanson model of ethylene (or another fragment) bonding in an organometallic molecule.[51] In the surface case, this is often termed the Blyholder model, the reference being to a perceptive early suggestion of such bonding for CO on surfaces.[41b] More generally, interactions ① and ② are the fundamental electronic origins of the cluster-surface analogy. This is a remarkably useful construction of a structural, spectroscopic, and thermodynamic link between organometallic chemistry and surface science. [52a]

For a beautiful comparison of structure, bonding, and reactivity in organometallic molecules and on surfaces, see the recent book by Albert and Yates. [52b]

BARRIERS TO CHEMISORPTION

The repulsive two-orbital, four-electron interaction that turns into an attractive bonding force when the electrons, rising in energy, are dumped at the Fermi level is not just a curiosity. I think that it is responsible for observed kinetic barriers to chemisorption and the possible existence of several independent potential energy minima as a molecule approaches a surface.

Consider a model molecule, simplified here to a single occupied level, approaching a surface. Some schematic level diagrams and an associated total energy curve are drawn in Fig. 33. The approach coordinate translates into electron interaction. Far away there is just repulsion, which grows as the

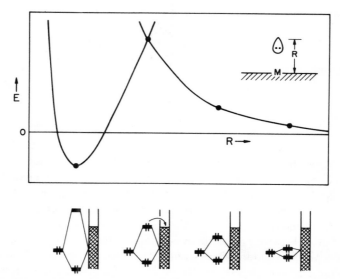

Figure 33 Schematic drawing showing how the interactions of levels (bottom) can lead to a potential energy curve (top) which has a substantial barrier to chemisorption. R measures the approach of a molecule, symbolized by a single interacting electron pair, to a surface. At large R repulsive four-electron interactions dominate. At some R (second point from left), the antibonding combination crosses the Fermi level and dumps its electrons. At shorter R there is bonding.

molecule approaches the surface. But when the antibonding combination is pushed up to the Fermi level, the electrons leave it for the reservoir of hole states, empty metal band levels. Further interaction is attractive.

To my knowledge, this simple picture was first presented by E. L. Garfunkel and C. Minot and their coworkers.[53] In reality, the repulsion at large metal adsorbate distances will be mitigated and, in some cases, overcome by attractive two-electron interactions of type ① or ② (see **54**). But the presence of the interaction, I think, is quite general. It is responsible, in my opinion, for some of the large kinetic barriers to CO dissociative chemisorption and CH_4 decomposition measured in the elegant beam experiments of S. T. Ceyer, R. J. Madix, and their coworkers.[54]

In reality, what we are describing is a surface crossing. And there may be not one, but several such, since it is not a single level but rather groups of levels that are "pushed" above the Fermi level. There may be several metastable minima, precursor states, as a molecule approaches a surface.[55]

In this section I have mentioned for the second time the bonding consequences of emptying, at the Fermi level, molecular orbitals delocalized over adsorbate and surface, and antibonding between the two. Salahub[50] and A. B. Anderson[56a] stress the same effect as, in another context, do Harris and Andersson. There is a close relationship between this phenomenon and a clever suggestion made some time ago by Mango and Schachtschneider[57] on the way in which metal atoms (with associated ligands) lower the activation barriers for forbidden concerted reactions. They pointed out that such electrons, instead of proceeding on to high antibonding levels, can be transferred to the metal. We, and others, have worked out the details of this kind of catalysis for some specific organometallic reactions, such as reductive elimination.[58] It's quite a general phenomenon, and we will return to it again in a subsequent section.

CHEMISORPTION IS A COMPROMISE

Consider again the basic molecule-surface interaction diagram **59**, now drawn specifying the bonding within each component. The occupied orbitals of the molecule A are generally bonding or nonbonding within that molecule, the unfilled orbitals of A are usually antibonding. The situation on the metal depends on where in a band the Fermi level lies: the bottom of the d band is metal–metal bonding, the top metal–metal antibonding. This is why the cohesive energy of the transition metals reaches a maximum around the middle of the transition series. Most of the metals of catalytic interest are in the middle or right part of the transition series. It follows that at the Fermi level the orbitals are generally metal–metal antibonding.

What is the effect of the various interactions on bonding within and between the adsorbate and the surface? Interactions ① and ② are easiest to analyze; they bind the molecule to the surface, and in the process they transfer electron density from generally bonding orbitals in one component to antibonding orbitals in the other. The net result: a bond is formed between the adsorbed molecule A and the surface. But bonding within the surface and within A is weakened, **60**.

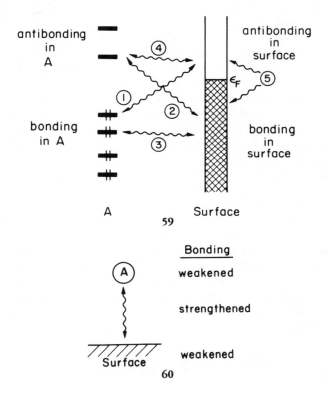

This is indicated schematically in **60**. What about interactions ③ and ④? For moderate interaction, ③ is repulsive and ④ has no effect. Neither does anything to bonding within A or the surface. When interaction grows, and antibonding (③) or bonding (④) states are swept past the Fermi level, these interactions provide molecule-surface bonding. At the same time, they weaken bonding in A, transferring electron density into antibonding levels and out of bonding ones. The effect of such strong interaction of type ③ or ④ or, more generally, of second order electron shifts, type ⑤, on bonding within the surface depends on the position of the Fermi level and the net electron drift.

The sum total of these interactions is still the picture of **60**: *metal-adsorbate bonding is accomplished at the expense of bonding within the*

Table 3 Bonding Characteristics of Several Acetylene Adsorption Sites on Pt(111)

	C_2H_2	Bare Surface				
Binding energy[a] (eV)			3.56	4.68	4.74	4.46
Overlap population						
C–C	1.70		1.41	1.32	1.21	1.08
Pt_1–Pt_2		0.14	0.12	0.08	0.09	− 0.02
Pt_2–Pt_3		0.14	0.14	0.13	0.07	0.06
Pt_1–Pt_4		0.14	0.13	0.13	0.15	0.06
Pt_1–C[b]			0.30	0.54	0.52	0.33
Pt_3–C			0.00	0.01	0.19	0.27
Occupations						
π^*	0.0		0.08	0.17	0.33	0.53
$\pi_\sigma{}^*$	0.0		0.81	1.06	1.03	0.89
π_σ	2.0		1.73	1.59	1.59	1.57
π	2.0		1.96	1.96	1.73	1.53

[a] Taken as difference: $E(slab) + E(C_2H_2) − E(geometry)$ in eV.
[b] The carbon atom here is the closest to the particular Pt atom under consideration.

metal and the adsorbed molecule. This is the compromise alluded to in the heading of this section.

A specific case will illustrate this point and show the way to an important consequence of this very simple notion. Earlier we drew four possible geometries for a layer of acetylene, coverage = 1/4, on top of Pt(111), **25**. Table 3 shows some of the indices of the interaction in the four alternative geometries, in particular the occupations of the four acetylene fragment orbitals (π, π_σ, π_σ^*, π^*), the various overlap populations, and calculated binding energies.

The threefold bridging geometry (**25c**) is favored, in agreement with experiment and other theoretical results.[29] One should say right away that this may be an accident—the extended Hückel method is not especially good at predicting binding energies. The twofold (**25b**) and fourfold (**25d**) sites are slightly less bound, but more stable than the onefold site, **25a**. But this order of stability is *not* a reflection of the extent of interaction. Let's see how and why this is so.

The magnitude of interaction could be gauged by looking at the acetylene fragment orbital populations, or the overlap population. In the detailed discussion of the twofold site in an earlier section, we saw π and π^* more or less unaffected, π_σ depopulated, π_σ^* occupied. As a consequence, Pt–C bonds are formed, the C–C bond is weakened, and (interaction ⑤) some Pt–Pt bonds on the surface are weakened. A glance at the fragment MO populations and overlap populations in Table 3[29] shows that all this happens much more in the fourfold site **25d**—note that even π and π^* get strongly involved. The most effective interaction here is that shown in **61**. Note that it is primarily of type ④.

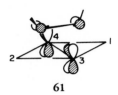

61

By any measure, interaction is least in the on-top or onefold geometry, most in the fourfold one. See, for instance, the trend in C–C overlap populations or the Pt–Pt bond weakening. In the fourfold geometry one Pt–Pt overlap population is even negative; bonding between metal atoms in the surface is being destroyed. It is clear that the favorable condition for chemisorption, or the preference of a hydrocarbon fragment for a specific surface site, is determined by a balance between increased surface-adsorbate bonding and loss of bonding within the surface or in the adsorbed molecules.

Adsorbate-induced surface reconstruction and dissociative chemisorption are merely natural extremes of this delicate balance. In each case, strong surface-adsorbate interactions direct the course of the transformation, either breaking up bonding in the surface, so that it reconstructs, or disrupting the adsorbed molecule.[59] An incisive discussion of the latter situation for the case of acetylene on iron and vanadium surfaces was provided by A. B. Anderson.[60]

FRONTIER ORBITALS IN THREE-DIMENSIONAL EXTENDED STRUCTURES

The frontier orbital way of thinking, especially with respect to donor-acceptor interactions, if of substantial utility in the solid state. Let me give one example here.

The Chevrel phases are a fascinating set of ternary molybdenum chalcogenide materials of varying dimensionality and interesting physical properties.[61] In the parent phase, epitomized by $PbMo_6S_8$, one has recognizable Mo_6S_8 clusters. In these clusters, shown in three views in **62**, sulfurs cap the eight faces of an octahedron of molybdenums. The Mo_6S_8 clusters are then embedded in a substructure of lead cubes (this is a thought construction of the structure!), as in **63**. But the structure doesn't remain here. In every cubical cell, the Mo_6S_8 rotates by ~26° around a cube diagonal, to reach structure **64** (Pb's are missing in this drawing, for clarity).

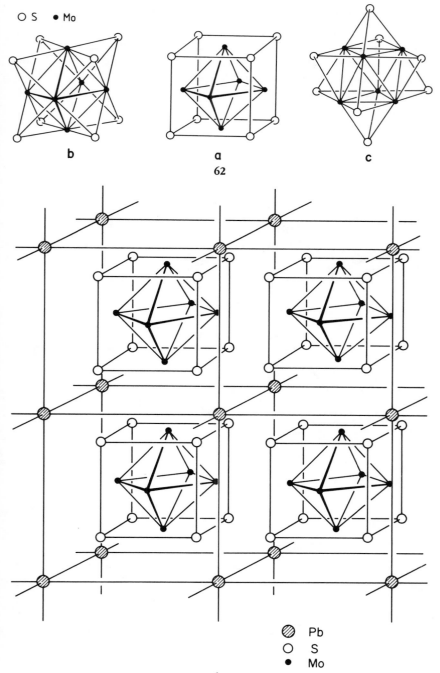

○ S   ● Mo

b   a   c

62

@ Pb
○ S
● Mo

63

Why? The answer is implicit in **64**. A rotation of roughly this magnitude is required to give each Mo within one unit a fifth bonding interaction with a sulfur of a cluster in the empty neighboring cube. If one does a molecular orbital calculation on the isolated cluster (Fig. 34), one finds that the five lowest empty orbitals of the cluster point out, away from the molybdenums, hungry for the electron density of a neighboring sulfur. [62,63a]

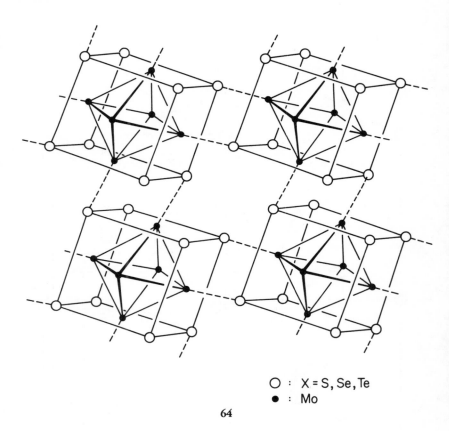

$\bigcirc$: X = S, Se, Te
● : Mo

64

The structure of this material is driven by donor-acceptor interactions. So it is for $In_3Mo_{15}Se_{19}$ and $K_2Mo_9S_{11}$, which contain $Mo_{12}X_{14}$ and Mo_9X_{11} clusters shown in **65**. [61] A molecular orbital calculation on each of these clusters shows prominent low-lying orbitals directed away from the terminal Mo's, just where the dashed lines are. [63a] That's how these clusters link and aggregate in their respective solid state structures.

This donor-acceptor analysis of the crystal structure indicates that if

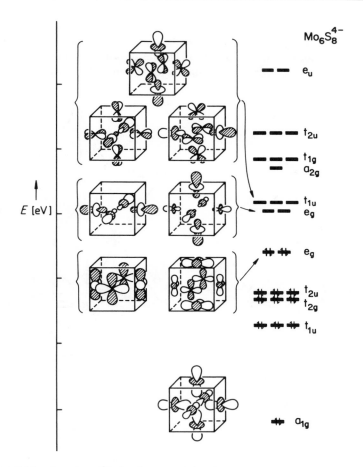

Figure 34 The frontier orbitals of an $Mo_6S_8^{4-}$ cluster, with some selected orbitals sketched. The lowest a_{1g} and the higher e_g and t_{1u} orbitals have substantial local z^2 character, i.e., point "out."

one wants to "solubilize" these clusters as discrete molecular entities, one must provide an alternative, better base than the molecule itself. Only then will one get discrete $Mo_6X_8 \cdot L_6{}^q$ complexes.[63b]

One more conclusion can be easily drawn from Fig. 34, one that applies what we know: when the clusters assemble into the lattice **64**, the five LUMOs of Fig. 34 will be pushed up by interactions with neighboring cube sulfurs. All the cluster levels will spread out into bands. Will the HOMO band be broad or narrow? That band is crucial because if you do the electron counting in $PbMo_6S_8$, you come to 22 electrons per Mo_6S_8, the top level in Fig. 34 half-filled. A glance at Fig. 34 shows that the level in question, of e_g symmetry, is made up of Mo d functions that are of δ type with respect to the Mo–S external axis. Bringing in the neighboring cells will provide little dispersion for this band. The result is a substantial DOS at the Fermi level, one of several requirements for superconductivity.[64]

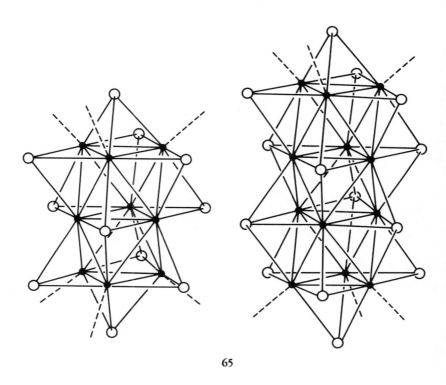

65

An interesting variation on the donor-acceptor theme in the solid is that the donor or acceptor need not be a discrete molecule, as one Mo_6S_8 cluster is toward another in the Chevrel phases. Instead, we can have electron transfer from one sublattice, one component of a structure, to another. We've already seen this in the explanation of the tuning of the X $\cdots$ X contact in the AB_2X_2 $ThCr_2Si_2$ structure. There the entire transition metal or B sublattice, made up of square nets, acts as a donor or acceptor, a reducing or oxidizing agent, for the X sublattice, made up of X $\cdots$ X pairs. A further example is provided by the remarkable $CaBe_2Ge_2$ structure, **66**.[65] In this structure, one B_2X_2 layer, **68**, has B and X components interchanging places relative to another layer, **67**. These layers are not identical, but isomeric. They will have different Fermi levels. One layer in the crystal will be a donor relative to the other. Can you reason out which will be the donor, which the acceptor layer? We will return to these molecules below.

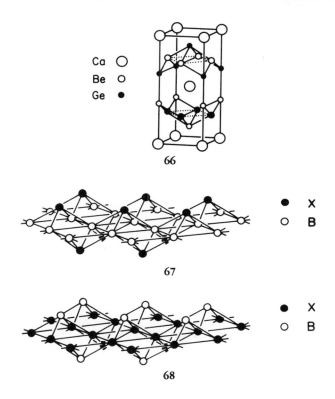

Ca ○
Be ◯
Ge ●

66

● X
○ B

67

● X
○ B

68

MORE THAN ONE ELECTRONIC UNIT IN THE UNIT CELL. FOLDING BANDS

Do you remember the beautiful platinocyanide stack? It has not yet exhausted its potential as a pedagogic tool. The oxidized platinocyanides are not eclipsed, **69a**, but staggered, **69b**. A polyene is not a simple linear chain, **70a**, but, of course, at least s-trans, zig-zag **70b**. Or it could be s–cis, **70c**. Obviously, there will be still other feasible arrangements; indeed, nature always seems to find one we haven't thought of.

In **69a** and **70a**, the unit cell contains one basic electronic unit, PtH_4^{2-}, a CH group. In **69b** and **70b**, the unit is doubled, approximately so in unit cell dimension, exactly so in chemical composition. In **70c**, we have four CH units per unit cell. A physicist might say that each is a case unto itself. A chemist is likely to say that probably not much has changed on doubling or quadrupling or multiplying by 17 the contents of a unit cell. If the geometric distortions of the basic electronic unit that is being repeated are not large, it is likely that any electronic characteristics of that unit are preserved.

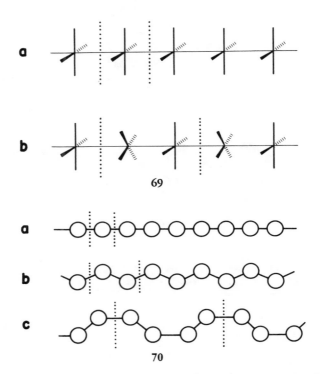

The number of bands in a band structure is equal to the number of molecular orbitals in the unit cell. So if the unit cell contains 17 times as many atoms as the basic unit, it will contain 17 times as many bands. The band structure may look messy. The chemist's feeling that the 17-mer is a small perturbation on the basic electronic unit can be used to simplify a complex calculation. Let's see how this goes, first for the polyene chain, then for the PtH_4^{2-} polymer.

Conformation **70a, b, c** differ from each other not just in the number of CH entities in the unit cell but also in their geometry. Let's take these one at a time. First prepare for the distortion from **70a** to **70b** by doubling the unit cell. Then, subsequently, distort. This sequence of actions is indicated in **71**.

Suppose we construct the orbitals of **71b**, the doubled unit cell polymer, by the standard prescription: (1) get MOs in unit cell, (2) form Bloch functions from them. Within the unit cell the MOs of the dimer are π and π^*, **72**. Each of these spreads out into a band, that of the π "running up," that of the π^* "running down," **73**. The orbitals are written out explicitly at the zone boundaries. This allows one to see that the top of the π band and the bottom of the π^* band, both at $k = \pi/2a$, are precisely degenerate. There is no bond alternation in this polyene (yet), and the two orbitals may have been constructed in a different way, but they obviously have the same nodal structure—one node every two centers.

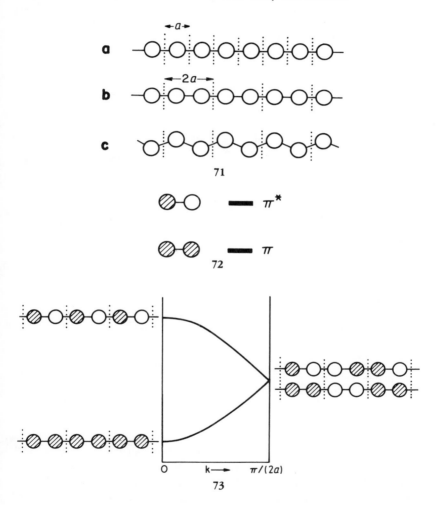

If we now detach ourselves from this viewpoint and go back and construct the orbitals of the one CH per unit cell linear chain **71a**, we get **74**. The Brillouin zone in **71b** (**73**) is half as long as it is here because the unit cell is twice as long.

At this point, we are hit by the realization that, of course, the orbitals of these polymers are the same. The polymers are identical; it is only some peculiar quirk that made us choose one CH unit as the unit cell in one case, two CH units in the other. I have presented the two constructions independently to make explicit the identity of the orbitals.

What we have is two ways of presenting the same orbitals. Band structure **73**, with two bands, is identical to **74**, with one band. All that has happened is that the band of the minimal polymer, one CH per unit cell, has been "folded back" in **74**. The process is shown in **75**.[66]

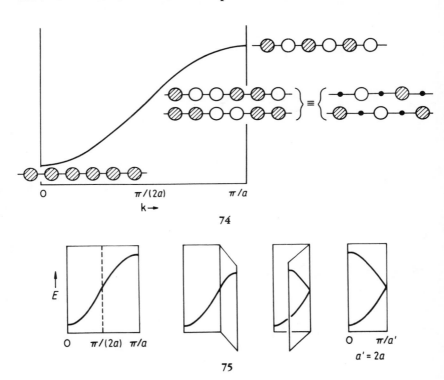

74

75

The process can be continued. If the unit cell is tripled, the band will fold as in **76a**. If it is quadrupled, we get **76b**, and so on. However, the point of all this is not just redundancy, seeing the same thing in different ways. There are two important consequences or utilizations of this folding. First, if a unit cell contains more than one electronic unit (and this happens often), then a realization of that fact, and the attendant multiplication of bands (remember **74** → **73** → **76a** → **76b**), allows a chemist to simplify the analysis in his or her mind. The multiplicity of bands is a consequence of an enlargement of the unit cell. By reversing, in our minds in a model calculation, the folding process by unfolding, we can go back to the most fundamental electronic act—the true monomer.

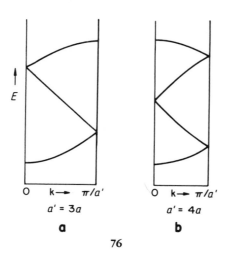

76

To illustrate this point, let me show the band structure of the *staggered* PtH_4^{2-} chain, **69b**. This is done in Fig. 35, left. There are twice as many bands in this region as there are in the case of the eclipsed monomer (the xy band is doubly degenerate). This is no surprise; the unit cell of the staggered polymer is $[PtH_4^{2-}]_2$. But it's possible to understand Fig. 35 as a small perturbation on the eclipsed polymer. Imagine the thought process **77a → b → c**, i.e., doubling the unit cell in an eclipsed polymer and then rotating every other unit by 45° around the z axis.

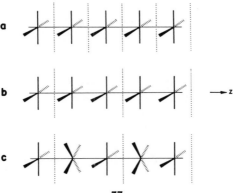

77

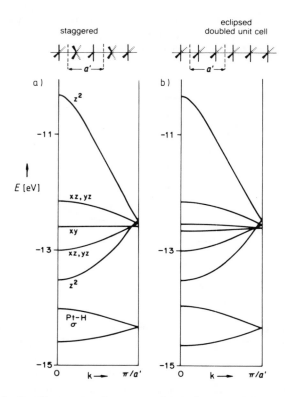

Figure 35 The band structure of a staggered PtH_4^{2-} stack (left), compared with the folded-back band structure of an eclipsed stack, two PtH_4^{2-} in a unit cell (right).

To go from **77a** to **b** is trivial, a simple folding back. The result is shown at the right of Fig. 35. The two sides of the figure are nearly identical. There is a small difference in the xy band, which is doubled, nondegenerate, in the folded-back eclipsed polymer (right-hand side of Fig. 13), but degenerate in the staggered polymer. What happened here could be stated in two ways, both the consequence of the fact that a real rotation intervenes between **77b** and **c**. From a group theoretical point of view, the staggered polymer has a new, higher symmetry element, an eightfold rotation-reflection axis. Higher symmetry means more degeneracies. It is easy to see that the two combinations, **78**, are degenerate.

Except for this minor wrinkle, the band structures of the folded-back eclipsed polymer and the staggered one are extremely similar. That allows us to reverse the argument, to *understand* the staggered one in terms of the eclipsed one plus the here minor perturbation of rotation of every second unit.

The chemist's intuition is that the eclipsed and staggered polymers *can't* be very different—at least, not until the ligands start bumping into each other, and for such steric effects there can be, in turn, much further intuition. The band structures may look different, since one polymer has one, the other two basic electronic units in the cell. Chemically, however, they should be similar, and we can see this by returning from reciprocal space to real space. Figure 36, which compares the DOS of the staggered and eclipsed polymers, shows just how alike they are in their distribution of levels.

There is another reason to feel at home with the folding process. The folding-back construction may be a prerequisite to understanding a chemically significant distortion of the polymer. To illustrate this point, we return to the polyene **71**. To go from **71a** (the linear chain, one CH per unit cell) to **71b** (linear chain, two CH's per unit cell) involves no distortion. However, **71b** is a way point, a preparation for a real distortion to the more realistic "kinked" chain, **71c**. It behooves us to analyze the process stepwise, **71a** → **71b** → **71c**, if we are to understand the levels of **71c**.

78

Of course, nothing much happens to the π system of the polymer on going from **71a**, **b** to **c**. If the nearest-neighbor distances are kept constant, then the first real change is in the 1, 3 interactions. These are unlikely to be large in a polyene, since the π overlap falls off very quickly past the bonding region. We can estimate what will happen by writing down some explicit points in the band, and deciding whether the 1, 3 interaction that is turned on is stabilizing or destabilizing. This is done in **79**. Of course, in a real CH polymer this kinking distortion is significant, but that has nothing to do with the π system, but rather is a result of strain.

However, there is another distortion that the polyene can and does undergo. This is double-bond localization, an example of the very important Peierls distortion, i.e., the solid state analogue of the Jahn–Teller effect.

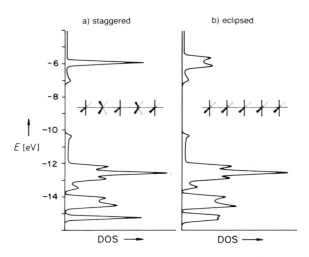

Figure 36 A comparison of the DOS of staggered (left) and eclipsed (right) $PtH_4{}^{2-}$ stacks.

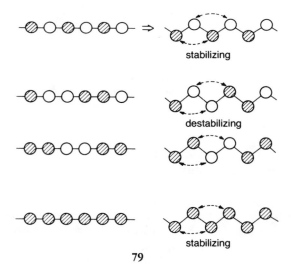

MAKING BONDS IN A CRYSTAL

When a chemist sees a molecular structure that contains several free radicals, orbitals with unpaired electrons, his or her inclination is to predict that such a structure will undergo a geometric change in which electrons will pair up, forming bonds. It is this reasoning, so obvious as to seem almost subconscious, that is behind the chemist's intuition that a chain of hydrogen atoms will collapse into a chain of hydrogen molecules.

If we translate that intuition into a molecular orbital picture, we have **80a**, a bunch (here six) of radicals forming bonds. That process of bond formation follows the H_2 paradigm, **80b**, i.e., in the process of making each bond a level goes down, a level goes up, and two electrons are stabilized by occupying the lower, bonding orbital.

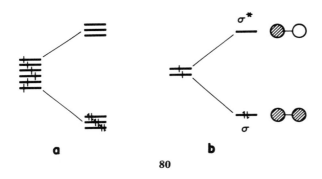

80

In solid state physics, bond formation has not had center stage, as it has in chemistry. The reasons for this are obvious: the most interesting developments in solid state physics have involved metals and alloys, and in these often close-packed or nearly close-packed substances, for the most part localized, chemical viewpoints have seemed irrelevant. For another large group of materials, ionic solids, it has also seemed useless to think of bonds. My contention is that there is a range of bonding—including what are usually called metallic, covalent, and ionic solids—and that there is, in fact, substantial overlap between seemingly divergent frameworks of bonding in these three types of crystals. I will take the view that the covalent approach is central and look for bonds when others wouldn't expect them. One reason for tolerating such foolhardiness might be that the other approaches (metallic, ionic) have had their day—why not give this one a chance? A second reason, one mentioned earlier, is that in thinking and talking about bonds in the crystal, one makes a psychologically valuable connection to molecular chemistry.

To return to our discussion of molecular and solid state bond formation, let's pursue the trivial chemical perspective of the beginning of this section. The guiding principle, implicit in **80**, is to *maximize bonding*. There may be impediments to bonding. One such impediment might be electron repulsion, another steric effects, i.e., the impossibility of two radicals to reach within bonding distance of each other. Obviously, the stable state is a compromise; some bonding may have to be weakened to strengthen some other bonding. But, in general, a system will distort so as to make bonds out of radical sites. Or, to translate this into the language of densities of states, *maximizing bonding in the solid state is connected to lowering the DOS at the Fermi level, moving bonding states to lower energy and antibonding ones to high energy.*

THE PEIERLS DISTORTION

In considerations of the solid state, a natural starting point is high symmetry—a linear chain, a cubic or close-packed three-dimensional lattice. The orbitals of the highly symmetrical, idealized structures are easy to obtain, but they do not correspond to situations of maximum bonding. These are less symmetrical deformations of the simplest, archetype structure.

The chemist's experience is usually the reverse, beginning from localized structures. However, there is one piece of experience we have that matches the thinking of the solid state physicist. This is the Jahn–Teller effect,[67] and it's worthwhile to show how it works by a simple example.

The Hückel π MOs of a square planar cyclobutadiene are well known. They are the one below two below one set shown in **81**. We have a typical Jahn–Teller situation, i.e., two electrons in a degenerate orbital. (Of course, we need worry about the various states that arise from this occupation, and the Jahn–Teller theorem really applies to only one.[67]) The Jahn–Teller theorem says that such a situation necessitates a large interaction of vibrational and electronic motion. It states that there must be at least one normal mode of vibration that will break the degeneracy and lower the energy of the system (and, of course, lower its symmetry). It even specifies which vibrations would accomplish this.

81

In the case at hand, the most effective normal mode is illustrated in **82**. It lowers the symmetry from D_{4h} to D_{2h}, and, to use chemical language, localizes double bonds.

The orbital workings of this Jahn–Teller distortion are easy to see. In **83**, Ψ_2 is stabilized: the 1–2, 3–4 interactions that were bonding in the square are increased; the 1–4, 2–3 interactions that were antibonding are decreased by the deformation. The reverse is true for Ψ_3—it is destabilized by the distortion at right. If we follow the opposite phase of the vibration, to the left in **82** or **83**, Ψ_3 is stabilized, Ψ_2 destabilized.

82

83

The essence of the Jahn–Teller theorem is revealed here: a symmetry-lowering deformation breaks an orbital degeneracy, stabilizing one orbital, destabilizing another. Note the phenomenological correspondence to **80** in the previous section.

One doesn't need a real degeneracy to benefit from this effect. Consider a nondegenerate two-level system, **84**, with the two levels of different symmetry (here labeled A, B) in one geometry. If a vibration lowers the symmetry so that these two levels transform as the same irreducible representation, call it C, then they will interact, mix, and repel each other. For two electrons, the system will be stabilized. The technical name of this effect is a second order Jahn–Teller deformation.[67]

The essence of the first or second order Jahn–Teller effect is that a high-symmetry geometry generates a real or near degeneracy, which can be broken with stabilization by a symmetry-lowering deformation. Note a further point: the level degeneracy is not enough by itself—one needs the right electron count. The cyclobutadiene (or any square) situation of **83** will be stabilized by a D_{2h} deformation for 3, 4, or 5 electrons, but not for 2 or 6 (S_4^{2+}).

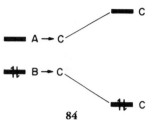

84

We can apply this framework to the solid. There is degeneracy and near degeneracy for any partially filled band. The degeneracy is that already mentioned, since $E(k) = E(-k)$ for any k in the zone. The near degeneracy is, of course, for k's just above or just below the specified Fermi level. For any such partially filled band there is available, in principle, a deformation that will lower the energy of the system. In the jargon of the trade, one says that the partial filling leads to an electron–phonon coupling that opens up a gap just at the Fermi level. This is the Peierls distortion,[68] the solid state counterpart of the Jahn–Teller effect.

Let's see how this works on a chain of hydrogen atoms (or a polyene). The original chain has one orbital per unit cell, **85a**, and an associated simple band. We prepare it for deformation by doubling the unit cell, **85b**. The band is typically folded. The Fermi level is halfway up the band; the band has room for two electrons per orbital, but for H or CH we have one electron per orbital.

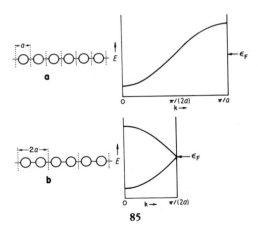

85

The phonon or lattice vibration mode that couples most effectively with the electronic motions is the symmetrical pairing vibration, **86**. Let's examine what it does to typical orbitals at the bottom, middle (Fermi level), and top of the band, **87**. At the bottom and top of the band nothing happens. What is gained (lost) in increased 1–2, 3–4, 5–6, etc., bonding (antibonding) is lost (gained) in decreased 2–3, 4–5, 6–7 bonding (antibonding). But in the middle of the band, at the Fermi level, the effects are dramatic. One of the degenerate levels there is stabilized by the distortion, the other destabilized. Note the phenomenological similarity to what happened for cyclobutadiene.

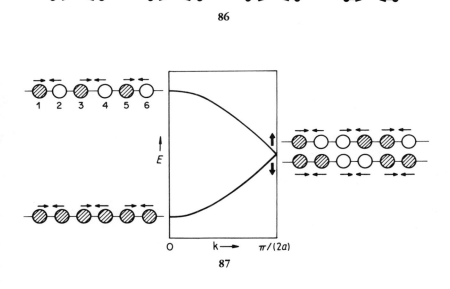

The action does not just take place at the Fermi level, but in a second order way the stabilization "penetrates" into the zone. It does fall off with k, a consequence of the way perturbation theory works. A schematic representation of what happens is shown in **88**. A net stabilization of the system occurs for any Fermi level, but obviously it is maximal for the half-filled band, and it is at that ϵ_F that the band gap is opened up. If we were to summarize what happens in block form, we'd get **89**. Note the resemblance to **80**.

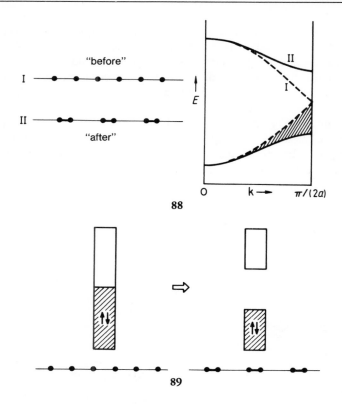

88

89

The polyene case (today it would be called polyacetylene) is especially interesting, for some years ago it occasioned a great deal of discussion. Would an infinite polyene localize, **90**? Eventually, Salem and Longuet-Higgins demonstrated that it would.[69] Polyacetylenes are an exciting field of modern research.[70] Pure polyacetylene is not a conductor. When it is doped, either partially filling the upper band in **89** or emptying the lower, it becomes a superb conductor.

90

There are many beautiful intricacies of the first and second order and low- or high-spin Peierls distortion, and for these the reader is referred to the very accessible review by Whangbo.[8]

The Peierls distortion plays a crucial role in determining the structure of solids in general. The one-dimensional pairing distortion is only one simple example of its workings. Let's move up in dimensionality.

One ubiquitous ternary structure is that of PbFCl (ZrSiS, BiOCl, Co_2Sb, Fe_2As). [16,71] We'll call it MAB here because in the phases of interest to us the first element is often a transition metal and the other components, A and B, are often main group elements. Diagram **91** shows one view of this structure, **92** another.

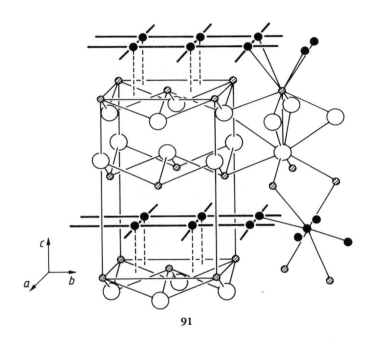

91

In the structure we see two associated square nets of M and B atoms, separated by a square net layer of A's. The A layer is twice as dense as the others, hence the MAB stoichiometry. Most interesting, from a Zintl viewpoint, is a consequence of that A layer density, a short A···A contact, typically 2.5 Å for Si. This is definitely in the range of some bonding. There are no short B···B contacts.

Some compounds in this series in fact retain this structure. Others distort, and it is easy to see why. Take GdPS. If we assign normal oxidation states of Gd^{3+} and S^{2-}, we come to a formal charge of P^- on the dense-packed P^- net. From a Zintl viewpoint, P^- is like S and so should form two bonds per P. This is exactly what it does. The GdPS structure[72] is shown in **93**, which is drawn after the beautiful representation of Hulliger et al.[72] Note the P–P cis chains in this elegant structure.

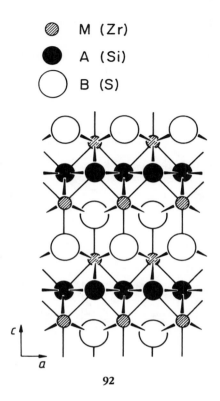

92

From the point of view of a band structure calculation, one might also expect bond formation, a distortion of the square net. Diagram **94** shows a qualitative DOS diagram for GdPS. What goes into the construction of this diagram is a judgment as to the electronegativities of Gd < P < S and the structural information that there are short P···P interactions in the undistorted square net, but no short S···S contacts. With the normal oxidation states of Gd^{3+}, S^{2-}, one comes to P^{-}, as stated above. This means that the P 3p band is two-thirds filled. The Fermi level is expected to fall in a region of a large DOS, as **94** shows. A distortion should follow.

The details of what actually happens are presented elsewhere.[16] The situation is intricate; the observed structure is only one of several likely ways for the parent structure to stabilize—there are others. Diagram **95** shows some possibilities suggested by Hulliger et al.[72] CeAsS chooses **95c**.[73] Nor is the range of geometric possibilities of the MAB phases exhausted by these. Other deformations are possible; many of them can be rationalized in terms of second order Peierls distortions in the solid.[16]

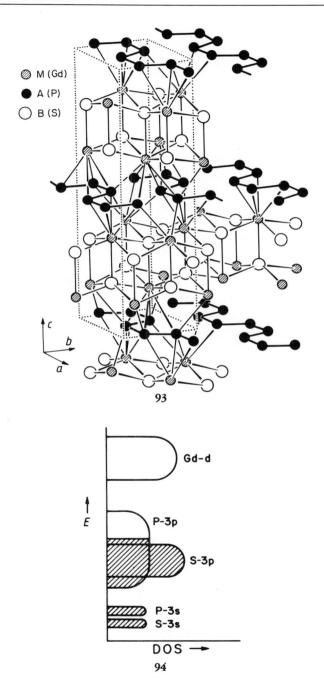

93

94

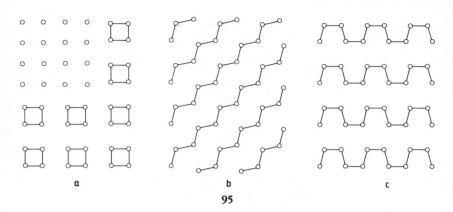

a   b   c

95

An interesting three-dimensional instance of a Peierls distortion at work (from one point of view) is the derivation of the observed structures of elemental arsenic and black phosphorus from a cubic lattice. This treatment is due to Burdett and coworkers.[6,74] The two structures are shown in their usual representation in **96**. It turns out that they can be easily related to a simple cubic structure, **97**.

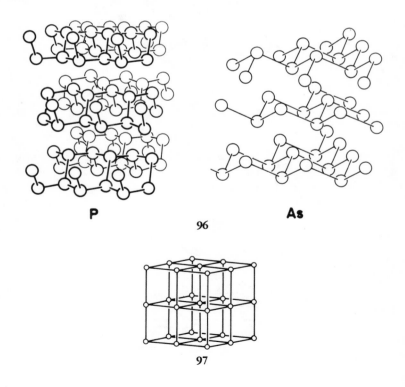

P   As

96

97

The DOS associated with the band structure of **97**, with one main group element of group 15 per lattice site, must have the block form **98**. There are five electrons per atom, so if the s band is completely filled, we have a half-filled p band. The detailed DOS is given elsewhere.[74] What is significant here is what we see without calculations, namely, a half-filled band. This system is a good candidate for a Peierls distortion. One pairing up all the atoms along x, y, and z directions will provide the maximum stabilization indicated schematically in **99**.

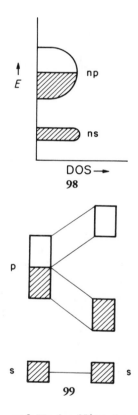

98

99

Burdett, McLarnan, and Haaland[74,a,c] showed that there are no less than 36 different ways to so distort. Two of these correspond to black phosphorus and arsenic, **100**. There are other possibilities as well.

There is one aspect of the outcome of a Peierls distortion—the creation of a gap at the Fermi level—that might be taken from the last case as being typical, but which is not necessarily so. In one dimension one can always find a Peierls distortion to create a gap. In three dimensions, atoms are much more tightly linked together. In some cases a stabilizing deformation leads to the formation of a real band gap, i.e., to an insulator or a semiconductor. In other cases, a deformation is effective in producing bonds, thereby pulling some states down from the Fermi level region. But because of the three-dimensional linkage it may not be possible to remove all the states from the Fermi level region. Some DOS remains there; the material may still be a conductor.

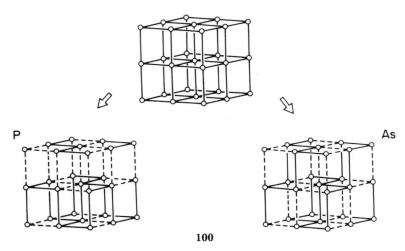

100

One final comment that is relevant to the $ThCr_2Si_2$ structure. The reader will note that we did not use a Peierls distortion argument in the resolution of the P–P pairing problem in that common structural type when we discussed it earlier. We could have done so, somewhat artificially, by choosing a structure in which the interlayer $P \cdots P$ separation was so large that the P–P σ and σ^* DOS came right at the Fermi level. Then a pairing distortion could have been invoked, yielding the observed bond. That, however, would have been a somewhat artificial approach. Peierls distortions are ubiquitous and important, but they're not the only way to approach bonds in the solid.

A BRIEF EXCURSION INTO THE THIRD DIMENSION

The applications discussed in the previous section make it clear that one must know, at least approximately, the band structure (and the consequent DOS) of two- and three-dimensional materials before one can make sense of their marvelous geometric richness. The band structures that we have discussed in detail have been mostly one- and two-dimensional.

Now let's look more carefully at what happens as we increase dimensionality.

Three dimensions really introduces little that is new, except for the complexities of drawing and the wonders of group theory in the 230 space groups. The s, p, d bands of a cubic lattice, or of face-centered or body-centered close-packed structures, are particularly easy to construct.[9,40] Let's look at a three-dimensional case of some complexity, the NiAs → MnP → NiP distortion.[75] The NiAs structure is one of the most common AB structures, with over a hundred well-characterized materials crystallizing in this type. The structure, shown in three different ways in **101**, consists of hexagonal close-packed layers that alternate metal and nonmetal atoms. To be specific, let's discuss the VS representative. The structure contains a hexagonal layer of vanadium atoms at $z = 0$, then a layer of sulfur atoms at $z = 1/4$, then a second layer of metal atoms at $z = 1/2$, superimposable on the one at $z = 0$, and, finally, a second layer of main group atoms at $z = 3/4$. The pattern is repeated along the c direction to generate a three-dimensional stacking of the type AbAcAbAc. It should not be imagined, however, that this is a layered compound; it is a tightly connected three-dimensional array. The axial V–V separation is 2.94 Å; the V–V contacts within the hexagonal net are longer, 3.33 Å.[75]

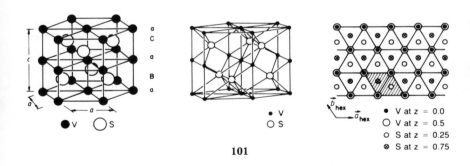

101

In terms of local coordination, each sulfur sits at the center of a trigonal prism of vanadiums, which in turn are octahedrally coordinated by six sulfurs. The V–S distances are typical of coordination compounds and, while there is no S–S bonding, the sulfurs are in contact with each other.

This is the structure of stoichiometric VS at high temperatures (>550°C). At room temperature, the structure is a lower symmetry, orthorhombic MnP one. The same structural transition is triggered by a subtle change in stoichiometry in VS_x, by lowering x from 1 at room temperature.[76]

The MnP structure is a small but significant perturbation on the NiAs type. Most (but not all) of the motion takes place in the plane perpendicular to the hexagonal axis. The net effect in each hexagonal net is to break it up into zig-zag chains, as in **102**. The isolation of the chains is exaggerated: the short V–V contact emphasized in **102** changes from 3.33 to 2.76, but the V–V distance perpendicular to the plane (not indicated in **102**) is not much longer (2.94 Å).

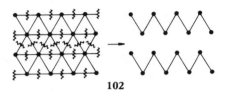

102

Still further distortions can take place. In NiP, the chains of Ni and P atoms discernible in the MnP structure break up into Ni_2 and P_2 pairs. For phosphides, it is experimentally clear that the number of available electrons tunes the transition from one structural type to another. Nine or 10 valence electrons favor the NiAs structure (for phosphides), 11–14 the MnP, and a greater number of electrons the NiP alternative. For the arsenides this trend is less clear.

The details of these fascinating transformations are given elsewhere.[75] It is clear that any discussion must begin with the band structure of the aristotype, NiAs (here computed for VS). This is presented in Fig. 37, which is a veritable spaghetti diagram, and seemingly beyond the powers of comprehension of any human being. Why not abdicate understanding, just let the computer spew these bands out and accept (or distrust) them? No, that's too easy a way out. We *can* understand much of this diagram.

First, the general aspect. The hexagonal unit cell is shown in **103**. It contains two formula units V_2S_2. That tells us immediately that we should expect $4 \times 2 = 8$ sulfur bands, two 3s separated from six 3p. And $9 \times 2 = 18$ vanadium bands, of which 10, the 3d block, should be lowest.

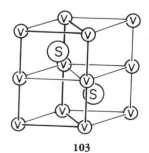

103

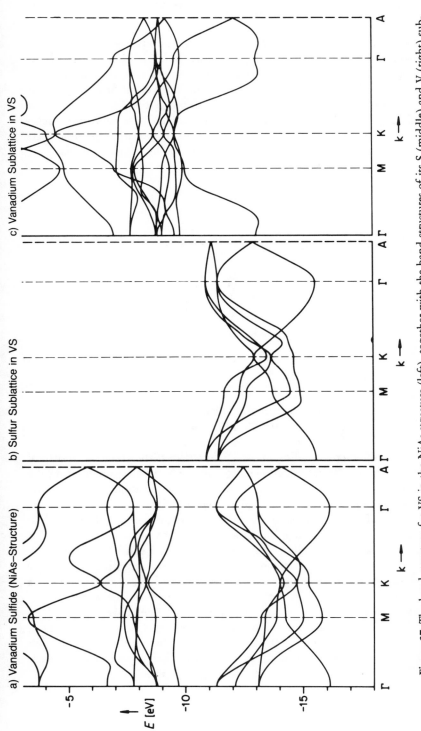

Figure 37 The band structure for VS in the NiAs structure (left), together with the band structures of its S (middle) and V (right) sublattices.

The Brillouin zone, **104**, has some special points labeled in it. There are conventions for this labeling.[9,15] The zone is, of course, three-dimensional. The band structure (Fig. 37) shows the evolution of the levels along several directions in the zone. Count the levels to confirm the presence of six low-lying bands (which a decomposition of the DOS shows to be mainly S 3p) and 10 V 3d bands. The two S 3s bands are below the energy window of the drawing. At some special points in the Brillouin zone there are degeneracies, so one should pick a general point to count bands.

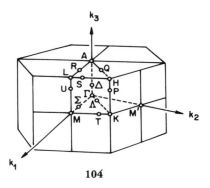

104

A feeling that this structure is made up of simpler components can be pursued by decomposing it into V and S sublattices. This is what Fig. 37b and c does. Note the relatively narrow V d bands around -8 to -9 eV. There is metal–metal bonding in the V sublattice, as shown by the widths of the V s, p bands. There are also changes in the V d bands on entering the composite VS lattice. A chemist would look for the local t_{2g}–e_g splitting characteristic of vanadium's octahedral environment.

Each of these component band structures could be understood in further detail.[77] Take the S 3p substructure at Γ. The unit cell contains two S atoms, redrawn in a two-dimensional slice of the lattice in **105** to emphasize the inversion symmetry. Diagrams **106–108** are representative x, y, and z combinations of one S two-dimensional hexagonal layer at Γ. Obviously, x and y are degenerate, and the x, y combination should be above z—the former is locally σ antibonding, the latter π bonding. Now combine two layers. The x, y layer Bloch functions will interact less (π overlap) than the z functions (σ antibonding for the Γ point, **109**). These qualitative considerations (x, y above z, the z bands split more than the x, y bands) are clearly visible in the positioning of bands 3–8 in Fig. 37a and b.

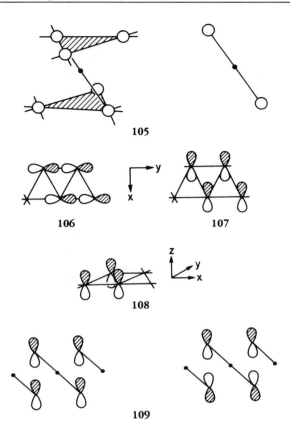

With more, admittedly tedious, work, every aspect of these spaghetti diagrams can be understood. And, much more interestingly, so can the electronic tuning of the NiAs → MnP → NiP displacive transition.[75]

Now let's return to some simpler matters, concerning surfaces.

QUALITATIVE REASONING ABOUT ORBITAL INTERACTIONS ON SURFACES

The previous sections have shown that one can work back from band structures and densities of states to local chemical actions—electron transfer and bond formation. It may still seem that the qualitative construction of surface-adsorbate or sublattice-sublattice orbital interaction diagrams in the forward direction is difficult. There are all these orbitals. How to estimate their relative interaction?

Symmetry and perturbation theory make such a forward construction relatively simple, as they do for molecules. First, in extended systems the wave vector k is also a symmetry label, classifying different irreducible representations of the translation group. In molecules, only levels belonging to the same irreducible representation interact. Similarly, in the solid only levels of the same k can mix with each other.[9,15]

Second, the strength of any interaction is measured by the same expression as for molecules:

$$\Delta E = \frac{|H_{ij}|^2}{E_i^0 - E_j^0}$$

Overlap and separation in energy matter, and can be estimated.[6,8,11]

There are some complicating consequences of there being a multitude of levels, to be sure. Instead of just saying that "this level does (or does not) interact with another one," we may have to say that "this level interacts more (or less) effectively with such and such part of a band." Let me illustrate this with some examples.

Consider the interaction of methyl, CH_3, with a surface, in on-top and bridging sites, **110**.[78] Let's assume low coverage. The important methyl orbital is obviously its nonbonding or radical orbital n, a hybrid pointing away from the CH_3 group. It will have the greatest overlap with any surface orbitals. The position of the n orbital in energy is probably just below the bottom of the metal d band. How to analyze the interactions of metal and methyl?

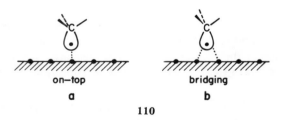

110

It's useful to take things apart and consider the metal levels one by one. Diagram **111** illustrates schematically some representative orbitals in the z^2 and xz bands. The orbitals at the bottom of a band are metal–metal bonding, those in the middle nonbonding, at the top antibonding. Although things are assuredly more complicated in three dimensions, these one-dimensional pictures are indicative of what transpires.

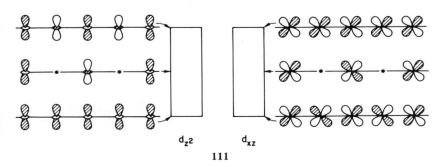

111

The methyl radical orbital (it's really a band, but the band is narrow for low coverage) interacts with the entire z^2 and xz bands of the metal, except at a few special symmetry-determined points where the overlap is zero. But it's easy to rank the magnitude of the overlaps, as I've done in 112 for on-top adsorption.

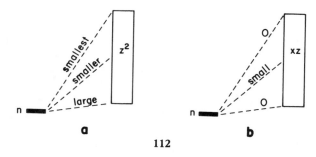

112

n interacts with the entire z^2 band but, because of the better energy match, more strongly so with the bottom of the band, as **113** shows. For interaction with xz, the overlap is zero at the top and bottom of the band, and never very efficient elsewhere, **114**. For adsorption in the bridge, as in **110b**, we would estimate the overlaps to go as **115**. There is nothing mysterious in these constructions. The use of the perturbation theoretical apparatus and specifically the role of k in delimiting interactions on surfaces goes back to the work of Grimley[45] and Gadzuk,[44] and has been consistently stressed by Salem.[47]

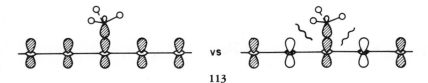

vs

113

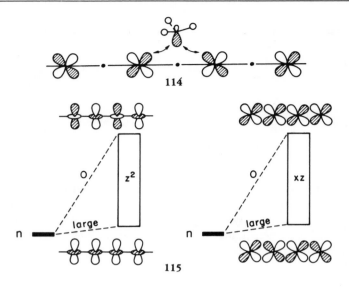

For a second example, let's return to acetylene on Pt(111), specifically in the twofold and fourfold geometries. [29] In the twofold geometry, we saw earlier (from the decomposition of the DOS) that the most important acetylene orbitals were π_σ and π_σ*. These point toward the surface. Not surprisingly, their major interaction is with the surface z^2 band. But π_σ and π_σ* interact preferentially with different parts of the band, picking out those metal surface orbitals which have nodal patterns similar to those of the adsorbate. Diagram 116 shows this; in the twofold geometry at hand the π_σ orbital interacts better with the bottom of the surface z^2 band and the π_σ* with the top of that band.

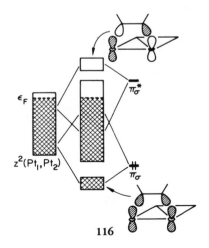

Note the "restructuring" of the z^2 band that results: in that band some metal–metal bonding levels that were at the bottom of the band are pushed up, while some of the metal–metal antibonding levels are pushed down. Here, very clearly, is part of the reason for weakening of metal–metal bonding on chemisorption.

We pointed out earlier that fourfold site chemisorption was particularly effective in weakening the surface bonding, and transferring electrons into π^* as well as π_σ^*, thus also weakening C–C bonding. The interaction responsible was drawn out in **61**. Note that it involves the overlap of π^* specifically with the top of the xz band. Two formally empty orbitals interact strongly, and their bonding component (which is antibonding within the metal and within the molecule) is occupied.

In general, it is possible to carry over frontier orbital arguments, the language of one-electron perturbation theory, to the analysis of surfaces.

THE FERMI LEVEL MATTERS

Ultimately one wants to understand the catalytic reactivity of metal surfaces. What we have learned experimentally is that reactivity depends in interesting ways on the metal, on the surface exposed, on the impurities or coadsorbates on that surface, on defects, and on the coverage of the surface. Theory is quite far behind in making sense of these determining factors of surface reactivity, but some pieces of understanding emerge. One such factor is the role of the Fermi level.

The Fermi level in all transition series falls in the d band—if there is a total of x electrons in the (n)d and (n + 1)s levels, then a not-bad approximation to the configuration or effective valence state of any metal is $d^{x-1}s^1$. The filling of the d band increases as one goes to the right in the transition series. But what about the position of the Fermi level?

What actually happens is shown schematically in **117** (a repeat of **48**), perhaps the most important diagram of metal physics. For a detailed discussion of the band structure, the reader is directed to the definitive work of O. K. Andersen. [40] Roughly, what transpires is that the center of gravity of the d band falls as one moves to the right in the transition series. This is a consequence of the ineffective shielding of the nucleus for one d electron by all the other d electrons. The magnitude of the ionization potential of a single d electron increases to the right. The orbitals also become more contracted, resulting in a less dispersed band as one move to the right. At the same time, the band filling increases. The position of the band center of gravity and the filling compete, the former wins out. Thus the Fermi level falls at the right side of the transition series. What happens in the middle is a little more complicated. [40]

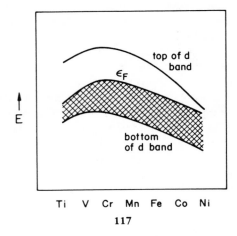

$$\uparrow E$$

117

Let's see the consequences of this trend for two chemical reactions. One is well studied, the dissociative chemisorption of CO. The other is less well known, but it certainly matters, for it must occur in Fischer–Tropsch catalysis. This is the coupling of two alkyl groups on a surface to give an alkane.

In general, early and middle transition metals break up carbon monoxide; late ones just bind it molecularly. [79] *How* the CO is broken up, in detail, is not known experimentally. Obviously, at some point the oxygen end of the molecule must come in contact with the metal atoms, even though the common coordination mode on surfaces, as in molecular complexes, is through the carbon. In the context of pathways of dissociation, the recent discovery of CO lying down on some surfaces, **118**, is intriguing. [80] Perhaps such geometries intervene on the way to splitting the diatomic to chemisorbed atoms. There is a good theoretical model for CO bonding and dissociation. [81]

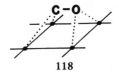

118

Parenthetically, the discovery of **118**, and of some other surface species bound in ways no molecular complex shows, should make inorganic and organometallic chemists read the surface literature not only to find references with which to decorate grant applications. The surface-cluster analogy, of course, is a two-way street. So far, it has been used largely to

Table 4 Some Orbital Populations in CO Chemisorbed on First Transition Series Surfaces (from Ref. 27)

	Electron Densities in Fragment Orbitals					
	Ti(0001)	Cr(110)	Fe(110)	Co(0001)	Ni(100)	Ni(111)
5σ	1.73	1.67	1.62	1.60	1.60	1.59
$2\pi^*$	1.61	0.74	0.54	0.43	0.39	0.40

provide information (or comfort for speculations) for surface studies, drawing on known molecular inorganic examples of binding of small molecules. But now surface structural studies are better, and cases are emerging of entirely novel surface-binding modes. Can one design molecular complexes inspired by structures such as **118**?

Returning to the problem of the metal surface influence on the dissociation of CO, we can look at molecular chemisorption, C end bonded, and see if there are any clues. Table 4 shows one symptom of the bonding on several different surfaces, the population of CO 5σ and $2\pi^*$.[27]

The population of 5σ is almost constant, rising slowly as one moves from the right to the middle. The population of $2\pi^*$, however, rises sharply. Not much is left of the CO bond by the time one gets to Ti. If one were to couple, dynamically, further geometric changes—allowing the CO to stretch, tilt toward the surface, and so on—one would surely get dissociation on the left side of the series.

The reason for these bonding trends is obvious. Diagram **119** superimposes the position of CO 5σ and $2\pi^*$ levels with the metal d band. 5σ will interact more weakly as one moves to the left, but the dramatic effect is on $2\pi^*$. At the right it interacts, as is required for chemisorption. But $2\pi^*$ lies above the d band. In the middle and left of the transition series, the Fermi level rises above $2\pi^*$. $2\pi^*$ interacts more, is occupied to a greater extent. This is the initial indicator of CO disruption.[27]

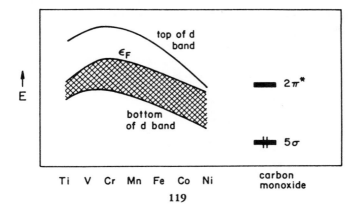

119

The second case we studied is one specific reaction likely to be important in the reductive oligomerization of carbon monoxide over a heterogeneous catalyst, the Fischer–Tropsch synthesis. The reaction is complicated and many mechanisms have been suggested. In the one I think likely, the carbide/methylene mechanism,[82] one follows a sequence of breaking up CO and H_2, and then hydrogenating the carbon to produce methyl, methylene, methyne on the surface, followed by various chain-forming associations of these and terminating reductive eliminations. It is one of those terminal steps that I want to discuss here, i.e., a prototype associative coupling of two adsorbed methyls to give ethane, **120**.[78]

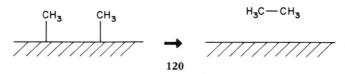

120

It's simple to draw **120**, but it hides a wonderful variety of processes. These are schematically dissected in **121**. First, given a surface and a coverage, there is a preferred site that methyls occupy, perhaps an equilibrium between several sites. Second, these methyls must migrate over the surface so as to come near each other. A barrier (call it the "migration energy") may intervene. Third, one methyl coming into the neighborhood of another may not be enough. It may have to come really close, e.g., on top of a neighboring metal atom. That may cost energy, for one is creating locally a high-coverage situation, one so high that it might normally be inaccessible. One could call this a steric effect, but let's call it a "proximity energy." Fourth, there is the activation energy to the actual C–C bond formation once the components are in place. Let's call this the "coupling barrier." Fifth, there might be an energy binding the product molecule to the surface. It is unlikely to be important for ethane, but might be substantial for other molecules.[83a] It is artificial to dissect the reaction in this way; nature does it all at once. But in our poor approach to reality (and here we are thinking in terms of static energy surfaces; we haven't even begun to do dynamics, to allow molecules to move on these surfaces), we can think of the components of the barrier impeding coupling: *migration + proximity + coupling + desorption energies.*

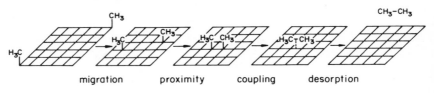

migration   proximity   coupling   desorption

121

To be specific, let's choose three dense surfaces: Ti(0001), Cr(110), and Co(0001). The calculations we carried out were for a three-layer slab, and initially a coverage of one-third. Three binding sites that were considered were on top or onefold, **122**, bridging or twofold, **123**, capping or threefold, **124**. The preferred site for each metal is the on-top site, **122**. [78]

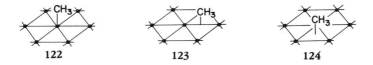

122   **123**   **124**

The total binding energy is greater for Ti than Cr than Co. Diagram **125** is an interaction diagram for CH_3 chemisorption. The CH_3 frontier orbital, a carbon-based directed radical lobe, interacts with metal s and z^2, much like the CO 5σ. Some z^2 states are pushed up above the Fermi level, and this is one component of the bonding. The other is an electron transfer factor. We started with a neutral surface and a neutral methyl. But the methyl lobe has room for two electrons. Metal electrons readily occupy it. This provides an additional binding energy. And because the Fermi levels increase to the left in the transition series, this "ionic" component contributes more for Ti than for Co. [78]

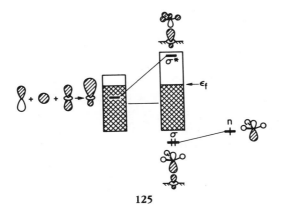

125

In a sense, these binding energies of a single ligand are not relevant to the estimation of relative coupling rates of two ligands on different surfaces. But even they show the effect of the Fermi level. A first step in coupling methyls is to consider the migration barriers of isolated groups. This is done in **126**. The relative energy zero in each case is the most stable on-top geometry.

Relative E (eV) :				
Co	:	0.0	1.1	1.4
Cr	:	0.0	0.9	0.9
Ti	:	0.0	0.5	0.1

126

The implication of **126** is that for Co the preferred migration itineraries are via bridged transition states, **127a**, for Ti via capping or hollow sites (**127b**), whereas for Cr both are competitive. For the reasons behind the magnitudes of the computed barriers, the reader is referred to our full paper.[78] Could one design an experiment to probe these migration alternatives? CH_3 has finally been observed on surfaces but remains a relatively uncommon surface species.[54]

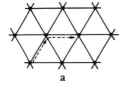   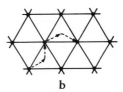

a   b

127

If we bring two methyl groups to on-top sites on adjacent metals, we see a splitting in the occupied CH_3 states. This is a typical two-orbital four-electron interaction, the way steric effects manifest themselves in one-electron calculations. If we compare the binding energy per methyl group in these proximate structures to the same energy for low-coverage isolated methyls, we get the calculated proximity energies of **128**. The destabilization increases with d electron count because some of the d levels occupied carry CH_3 lone pair contributions.

What happens when two CH_3 groups actually couple? The reaction begins with both CH_3 lone pairs nearly filled, i.e., a representation near CH_3^-. A new C–C σ bond forms and, as usual, we must consider σ and σ^* combinations, $n_1 \pm n_2$. Both are filled initially, but as the C–C bond forms, the σ^* combination will be pushed up. Eventually, it will dump its electrons into the metal d band.

Co	0.7 eV
Cr	0.5
Ti	-0.1

128

The actual evolution of the DOS and COOP curves allows one to follow this process in detail. For instance, Fig. 38 shows the contribution of the methyl n orbital, the radical lobe, to the total DOS along a hypothetical coupling reaction coordinate. Note the gradual formation of a two-peaked structure. COOP curves show the lower peak is C–C bonding, the upper one C–C antibonding. These are the σ and σ^* bonds of the ethane that is being formed.

The total energy of the system increases along the reaction path, as $n_1 - n_2$ becomes more antibonding. At the Fermi level, there is a turning point in the total energy. $\sigma^* = n_1 - n_2$ is vacated. The energy decreases, following $\sigma = n_1 + n_2$. The position of the Fermi level determines the turning point. So the coupling activation energy is expected to be greater for Ti than for Cr than for Co because, as noted above, the Fermi level is higher for the early transition metals, despite the lower d electron count. The reader familiar with reductive eliminations in organometallic chemistry will note essential similarities.[58,83b] We mention here again the relationship of our argument to the qualitative notions of Mango and Schachtschneider regarding how coordinated metal atoms affect organic reactions.[57]

The position of the Fermi level and the nature of the states at that level clearly is an important factor in determining binding and reactivity on metal surfaces. The point is not original to this work, but has been clearly discussed in several contributions to the literature.[56b,83c] Attention is directed to a particularly interesting discussion of how the local DOS at the Fermi level is affected by chemisorption.[83d]

ANOTHER METHODOLOGY AND SOME CREDITS

There have been an extraordinary number of theoretical contributions to solid state and surface science.[84] These have come from physicists and chemists, and have ranged from semiempirical molecular orbital (MO)

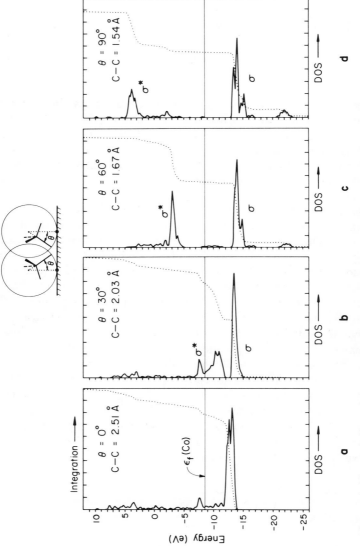

Figure 38 The evolution of the contribution of methyl lone pairs to the DOS of a chemisorption system (CH_3) on Co(0001) as the two methyls couple to give ethane. θ is defined at top. Note the development of two peaks corresponding to σ and σ^* of the CC bond in ethane.

calculations to state-of-the-art Hartree–Fock self-consistent-field plus configuration interaction (CI) and advanced density-functional procedures. Some people have used atom and cluster models, some extended slab or film models for surfaces. I will not discuss all these contributions, even those relevant to the systems I've mentioned, because (1) this is not an exhaustive review of theoretical methods, (2) I'm lazy, and (3) the field is full of conflicting claims of validity for the theoretical methods used. Such claims are of course typical of the reality (rather than the ideology) of all science. But theory is especially prone to them—because theorists rarely deal with the material world, but mostly with the abstract, bordering on the spiritual. That's inherent in the nature of answering the questions, the necessary and deep questions, "why" and "how". Basically, I'm not sure I *can* answer the question of whether one or another method is better, nor do I have the courage to try.

Most of the theoretical methods at hand are just better ways of solving the wave equation for the complex system at hand, not necessarily leading to more chemical and physical understanding. There is one exception, the complex of ideas on chemisorption introduced and developed by Lundqvist, Nørskov, Lang, and their coworkers. [85] This is a methodology rich in physical understanding, and because of that and the fact that it provides a different way of looking at barriers for chemisorption, I want to mention the method explicitly here.

The methodology focuses, as many density-functional schemes do, on the key role of the electron density. The Schrödinger equation is then solved self-consistently in the Kohn–Sham scheme. [86] Initial approaches dealt with a jellium-adatom system, which would at first sight seem rather unchemical, lacking microscopic detail. But there is much physics in such an effective medium theory, and with time the atomic details at the surface have come to be modeled with greater accuracy.

An example of the information the method yields is shown in Fig. 39, the total energy and density of states profile for H_2 dissociation on Mg(0001). [87] There are physisorption (P), molecular chemisorption (M), and dissociative chemisorption wells, with barriers in between. The primary controlling factor in molecular chemisorption is increasing occupation of H_2 $\sigma_u{}^*$, whose main density of states drops to the Fermi level and below as the H_2 nears the surface.

In this and other studies by this method, one can see molecular levels, sometimes spread into bands, moving about in energy space. But the motions seem to be different from those calculated by the extended Hückel procedure. Figure 18 showed for H_2 on Ni[88] some $\sigma_u{}^*$ density coming below the Fermi level. But the main peak of $\sigma_u{}^*$ was pushed up, as a simple interaction diagram might suggest, but in apparent disagreement with the result of Fig. 39. Perhaps (I'm not sure) one way to reconcile the two pictures

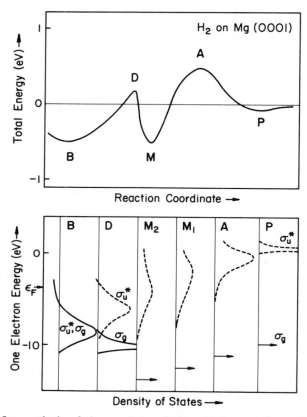

Figure 39 Some calculated characteristics of H_2 on Mg(0001), after Ref. 87. Top: schematic potential energy curve. P = physisorption minimum; M = chemisorbed molecule; B = chemisorbed atoms; A and D are transition states for chemisorption and dissociation. Bottom: development of the one-electron density of states at certain characteristic points. M_1 and M_2 correspond to two molecular chemisorption points, different distances from the surface. The dashed line is the σ_u^* density, moving to lower energy as the dissociation proceeds.

is by recognizing that mine is not self-consistent, i.e., does not account for proper screening of H_2 as it approaches the surface. It is possible that if self-consistency or screening by electrons in the metal were included in the one-electron formalism that the pictures could be reconciled. Also there is less discrepancy between the two approaches than one might imagine. In the reaction coordinate of Fig. 39 the H–H bond is stretched along the progression $P \rightarrow A \rightarrow M_1 \rightarrow M_2 \rightarrow D \rightarrow B$. σ_u^* drops precipitously, in our calculations as well, as the H–H bond is stretched.

The barriers to chemisorption in the work of Nørskov et al.[87] come from the initial dominance of "kinetic energy repulsion." This is the Pauli effect at work, and I would like to draw a correspondence between our four-electron repulsion and this kinetic energy effect. The problem (as usual) is that different models build in different parts of physical reality. It becomes very difficult to compare them. The reason the effort is worth making is that the Lundqvist–Nørskov–Lang model has proven remarkably useful in revealing trends in chemisorption. It is physically and chemically appealing.

There are some contributors to theoretical solid state chemistry and surface science whom I should like to mention because of their special chemical orientation. One is Alfred B. Anderson, who analyzed most important catalytic reactions, anticipating many of the results on surfaces presented in this book.[89] Evgeny Shustorovich and Roger C. Baetzold, working separately and together, both carried out detailed calculations of surface reactions and came up with an important perturbation theory-based model for chemisorption phenomena.[90] Christian Minot worked out some interesting chemisorption problems.[91] Myung-Hwan Whangbo's analyses of the bonding in low-dimensional materials such as the niobium selenides, TTF-type organic conductors, and molybdenum bronzes, as well as his recent studies of the high-T_c superconductors, contributed much to our knowledge of the balance of delocalization and electron repulsion in conducting solids.[7,8] Jeremy Burdett is responsible for the first new ideas on what determines solid state structures since the pioneering contributions of Pauling.[5,6] His work is consistently ingenious and innovative.[93]

Not the least reason I mention these people is pride: all of them have at some time (prior to doing their important independent work) visited my research group.

WHAT'S NEW IN THE SOLID?

If all the bands in a crystal are narrow (as they are in molecular and most extremely ionic solids), i.e., if there is little overlap between repeating molecular units, then there is no new bonding to speak of. But if at least some of the bands are wide, then there is delocalization, new bonding, and a molecular orbital picture is necessary. This is not to say that we cannot recover, even in such a large-dispersion, delocalized situation, local bonding. The preceding sections have shown that we can see bonds. But there may be qualitatively new bonding schemes that result from substantial delocalization. Recall in organic chemistry the consequences of aromaticity, and in inorganic cluster chemistry of skeletal electron pair counting algorithms.[21] In the beginning of this final section, I would like to trace some of the novel features of bonding in extended systems.

The language of orbital interactions and perturbation theory provides a tool that is applicable to the analysis of these highly delocalized systems, just as it works for small, discrete molecules. For instance, take the question posed at the end of a previous section. We have two isomeric two-dimensional lattices 67 and 68. Which will be a donor relative to the other? And which will be most stable?

These lattices are built up from two elements B and X in equal numbers, occupying two sublattices, I and II in 129. The elements are of unequal electronegativity, in the general case. In $ThCr_2Si_2$ one is a transition metal, the other a main group element, in $CaBe_2Ge_2$ each a main group element. Let's take, for purposes of discussion, the latter case as a model and write an interaction diagram for what happens locally, 130.

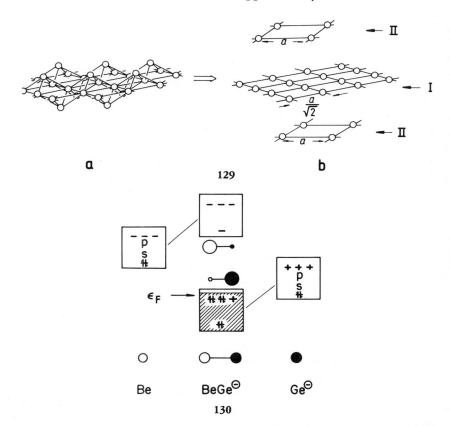

129

130

The diagram has been drawn in such a way that the more electronegative element is X. No implication as to bandwidth is yet made—the orbital blocks are just that, blocks, indicating the rough position of the levels. The lower block of levels is obviously derived from or localized in the orbitals of the more electronegative (here X) element. The band filling is actually appropriate to the CaBe$_2$Ge$_2$ structure, i.e., Be$_2$Ge$_2{}^{2-}$, or BeGe$^-$, or seven electrons per two main group atoms.

The orbitals develop into bands. The width of the bands depends on the inter-unit-cell overlap. The site II atoms are much farther apart from each other than the site I atoms (recall here the short metal–metal contacts in the ThCr$_2$Si$_2$ structure). We can say that sublattice I is more *dispersive* than sublattice II. The orbitals of atoms placed in sublattice I will form wider bands than those in sublattice II.

Now we have two choices: the more electronegative atoms can enter the less dispersive sites (lattice II) or the more dispersive sites (lattice I). The consequences are shown in **131** and **132**.

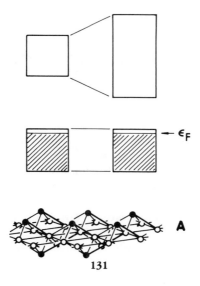

131

Which layer will be most stable and which will have the higher Fermi level depends on the electron filling. For a case such as CaBe$_2$Ge$_2$, or in general where the lower band is more than half-filled, the more electronegative atom will prefer the less dispersive site **131** and that layer will have a higher ionization potential, be a poorer donor.

The stability conclusion bears a little elaboration. It is based on the same "overlap repulsion" argument that was behind the asymmetrical splitting of hydrogen chain bands (Fig. 1). When orbitals interact, the antibonding combinations are more antibonding than the bonding ones are bonding. Filling antibonding combinations, filling the tops of widely dispersed bands, is costly in energy. Conclusions on stability, as is the case in molecular chemistry, depend strongly on the electron count. In this particular case, if the lower band were less than half-filled, the conclusion would be reversed, the more electronegative element would prefer the more dispersive site.

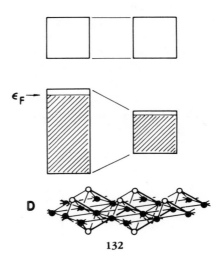

132

For ThCr$_2$Si$_2$ AB$_2$X$_2$ structures the conclusion we reach, that the more electronegative element should enter the less dispersive site, implies that for most cases the main group X component will prefer the less dispersive, square pyramidal, sublattice II positions. In CaBe$_2$Ge$_2$, Ge is more electronegative than Be. That means the layer in which the Ge enters the more dispersive sites (the bottom layer in **66**) should be a donor relative to the upper layer.

A reasonable question to ask is the following. If one layer (the acceptor layer) in CaBe$_2$Ge$_2$ is more stable than the other, the donor layer, why does the CaBe$_2$Ge$_2$ structure form at all? Why doesn't it go into a ThCr$_2$Si$_2$ structure based on the acceptor layer alone? The answer lies in the balance of covalent and dative interactions; for some elements the binding energy gained in donor-acceptor *interlayer* interactions overcomes the inherent stability of one layer isolated. [39c]

At times the perturbation introduced by delocalization may be strong enough to upset the local, more "chemical" bonding schemes. Let me sketch two examples here.

Marcasites and arsenopyrites are a common structural choice for MA$_2$ compounds, where M is a late transition metal, A a group 15 or 16 element. The structure, **133**, is related to the rutile one, in that one can easily perceive in the structure the octahedral coordination of the metal, and one-dimensional chains of edge-sharing octahedra. The ligands are now interacting, however; not $2(O^{2-})$ as in rutile, but S_2^{2-} or P_2^{4-} diatomic units in the marcasites.[92]

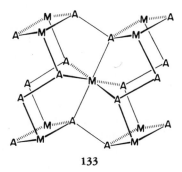

133

Low dimensionality characterizes another set of MS_2 sublattices, now in ternary structures of the type of $KFeS_2$ or $Na_3Fe_2S_4$.[93,94] In these molecules one finds one-dimensional MS_2 chains, consisting of edge-sharing tetrahedra, **134**.

134

In both of these structural types, characterized by their simplicity, the metal–metal separations are in the range 2.6–3.1 Å, where reasonable men or women might disagree whether there is much metal–metal bonding. Cases with bridging ligands are ones in which real metal–metal bonding is particularly difficult to sort out from bonding through the bridge. Certainly the metal–metal bonding doesn't look to be very strong, if it's there at all. So a chemist would start out from the local metal site environment, which is strikingly simple.

One would then predict a three below two orbital splitting at each metal in the octahedral marcasites and a two below three splitting in the tetrahedral MS_2 chains. The magic electron counts for a closed-shell, low-spin structure should be then d^6 for the octahedral **133**, d^4 for the tetrahedral **134**. Forming the one-dimensional chains and then the three-dimensional structure will introduce some dispersion into these bands, one might reason; but not much—appropriate electron counts for semiconducting or nonmagnetic behavior should remain d^6 for **133** and d^4 for **134**.

The experimental facts are as follows: the d^6 marcasites and arsenopyrites are semiconducting but, surprisingly, so are the d^4 ones. Most of the AMS_2 structures synthesized to date feature the metal atom in configurations between d^5 and $d^{6.5}$. The measured magnetic moments are anomalously low.

When calculations on these chains are carried out, one finds, to one's initial surprise, that the octahedral marcasite structure has a band gap at d^4 as well as d^6, and that the tetrahedral chain has a band gap at $d^{5.5}$ and not d^4. It seems that local crystal field considerations don't work. What in fact happens (and here the reader is referred to the detailed explanation in our papers[92,94]) is that the local field is a good starting point, but that further delocalizing interactions (and these are ligand–ligand and metal–ligand, and not so much metal–metal in the distance range considered) must be taken into account. The extended interactions modify the magic or gap electron counts that might be expected from just looking at the metal site symmetry.

In a preceding section, I outlined the orbital interactions that are operative in the solid state. These were the same ones as those that govern molecular geometries and reactivity. But there were some interesting differences, a consequence of one of the interacting components, the surface, having a continuum of levels available at the Fermi level. This provided a way to turn strong four- and zero-electron two-orbital interactions into bonding ones. As a corollary, there are shifts around the Fermi level that have bonding consequences. Let's look at this new aspect in detail in an example we had mentioned before, acetylene chemisorbed in low coverage in a parallel, twofold bridging mode on Pt(111), **135**.[29]

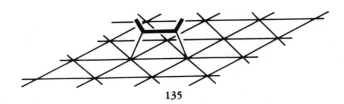

135

The most important two-electron bonding interactions that take place are between two of the acetylene π orbitals, π_σ and π_σ^* (see **136**), and the d band. π_σ and π_σ^* "point" toward the surface, have greater overlap with metal overlaps, and they interact preferentially with different parts of the band, picking out those metal surface orbitals with nodal patterns similar to that of the adsorbate. **137** shows this; in the "parallel bridging" geometry at hand the π_σ orbital interacts better with the bottom of the surface z^2 band and the π_σ^* with the top of that band.

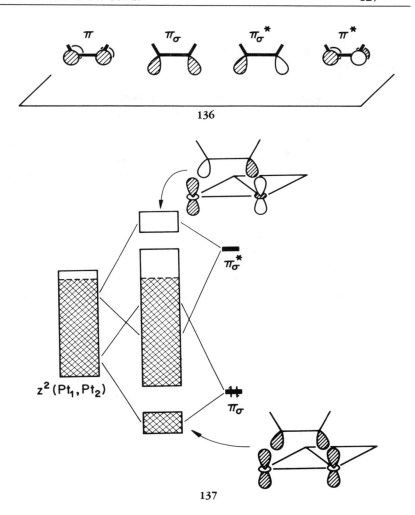

136

137

Both of these interactions are primarily of type ① and ③ (see **54** or **59**), four-electron repulsive or two-electron attractive interactions. Actually, the energetic and bonding consequences are a little complicated: the $z^2-\pi_\sigma$ interaction would be destabilizing if the antibonding component of this interaction remained filled, below the Fermi level. In fact, many $z^2-\pi_\sigma$ antibonding states are pushed above the Fermi level, vacated. This converts a destabilizing, four-electron interaction into a stabilizing two-electron one.

A counterpart to this interaction is ⑤. Normally we would not worry about zero-electron interactions because there is no "power" in them if there are no electrons. However, in the case of a metal with a continuous band of states, some of these levels—these are bonding combinations of π_σ^*

with the top of the z^2 band, as indicated in **137**—come below the Fermi level and are occupied. Therefore they also contribute to bonding the adsorbate to the surface.

It should be noted that a consequence of all these interactions is not only strengthening of metal–acetylene bonding, but also a weakening of bonding within the acetylene *and* within the metal. Interaction means delocalization, which in turn implies charge transfer. Interactions ① and ② operate to depopulate π_σ and populate $\pi_\sigma{}^*$, both actions weakening the acetylene π bond. Removing electrons from the bottom of the z^2 band, and better filling the top of that bond, both result in a weakening of the Pt–Pt bond.

Interaction ⑤, peculiar to the solid, is a reorganization of the states around the Fermi level as a consequence of primary interactions ①, ②, ③, and ④. Consider, for instance, the levels that are pushed up above the Fermi level as a result of interaction ③, the four-electron repulsion. One way to think about this is the following: the electrons do not, in fact, go up past the Fermi level (which remains approximately constant) but are dumped at the Fermi level into levels somewhere in the solid. This is shown schematically in **138**.

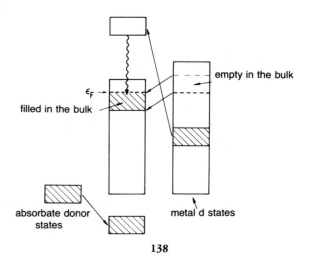

138

But where is "somewhere?" The electrons that come in come largely from regions that are not directly involved in the bonding with the adsorbate. In the case at hand, they may come from Pt bulk levels, from Pt surface atoms not involved with the acetylene, even from the Pt atoms binding the acetylene, but from orbitals of these atoms not used in that binding. While the metal surface is a reservoir of such electrons, these electrons are not innocent of bonding. They are near the top of their respective band, and as such are metal–metal antibonding. Thus interaction ⑤ weakens bonding in the surface. Together with the aforementioned electron transfer effects of interactions ①, ②, ③, and ④, it is responsible for adsorbate-induced surface reconstruction.

In general, as I already outlined in a previous section, nondissociative chemisorption is a delicate balance of the very same interactions, which weaken bonds in the adsorbed molecule and in the surface. Dissociative chemisorption and surface reconstruction are just two extremes of the same phenomenon.

So what's new in the solid? My straw-man physicist friend, thinking of superconductivity, charge and spin density waves, heavy fermions, solitons, nonlinear optical phenomena, ferromagnetism in its various guises—all the fascinating things of interest to him and that I've neglected—might say, "Everything." An exaggeration of what I've said in this book is, "Not much." There are interesting, novel consequences of delocalization and wide-band formation, but even these can be analyzed in the language of orbital interactions.

It would not surprise anyone if the truth were somewhere in between. It is certainly true that I've omitted, by and large, the origins of most of the physical properties of the solid, especially superconductivity and ferro-magnetism, which are peculiar to that state of matter. Chemists will have to learn much more solid state physics than I've taught here if they are to understand these observables, and they *must* understand them if they are to make rational syntheses.

What I have tried to do in this book and the published papers behind it is to move simultaneously in two directions—to form a link between chemistry and physics by introducing simple band structure perspectives into chemical thinking about surfaces. And I have tried to interpret these delocalized band structures from a very chemical point of view—via frontier orbital considerations based on interaction diagrams.

Ultimately, the treatment of electronic structure in extended systems is no more complicated (nor is it less so) than in discrete molecules. The bridge to local chemical action advocated here is through decompositions of the DOS and the crystal orbital overlap population (COOP) curves. These deal with the fundamental questions: Where are the electrons? Where can I find the bonds?

With these tools in hand, one can construct interaction diagrams for surface reactions, as one does for discrete molecules. One can also build the electronic structure of complicated three-dimensional solids from their sublattices. Many similarities between molecules and extended structures emerge, as do some novel effects that are the result of extensive delocalization.

I have concentrated on the most chemical notion of all, i.e., the solid is a molecule, a big one, to be sure, but just a molecule. Let's try to extract from the perforce delocalized picture of Bloch functions the chemical essence, the bonds that determine the structure and reactivity of this large molecule. The bonds must be there.

References

1. There is no single, comprehensive textbook of solid state chemistry. Perhaps this reflects the diversity of the field and the way it has grown from many different subdisciplines of chemistry, physics, and engineering. Among the books I would recommend are:
   (a) Wells, A. F., "Structural Inorganic Chemistry," 5th ed. Oxford University Press, Oxford, 1984.
   (b) Krebs, M. "Fundamentals of Inorganic Crystal Chemistry," McGraw-Hill, 1968.
   (c) West, A. R. "Solid State Chemistry and its Applications," John Wiley and Sons, New York, 1984.
   (d) O'Keeffe, M.; Navrotsky, A. (eds.). "Structure and Bonding in Crystals," Vols. 1 and 2, Academic Press, New York, 1981.
   (e) Cox, P. A. "The Electronic Structure and Chemistry of Solids," Oxford University Press, Oxford, 1981.

2. For a highly readable introduction to surface chemistry, see Somorjai, G. A. "Chemistry in Two Dimensions: Surfaces," Cornell University Press, Ithaca, 1981. See also Rhodin, T. N., Ertl, G. (eds.). "The Nature of the Surface Chemical Bond," North-Holland, Amsterdam, 1979.

3. (a) Zintl, E.; Woltersdorf, G. Z. Elektrochem., 1935, 41, 876; Zintl, E.; Angew. Chem. 1, 52 (1939).
   (b) Klemm, W.; Busmann, E. Z. Anorg. Allgem. Chem. 1963, 319, 297; (1963). Klemm, W. Proc. Chem. Soc. London 329 (1958).
   (c) Schäfer, H.; Eisenmann, B.; Müller, W. Angew. Chem. 85, 742 (1973). Angew Chem. Int. Ed. 12, 694 (1973).
   (d) Schäfer, H. Ann. Rev. Mater. Sci. 15, 1 (1985) and references therein.

4. Hoffmann, R. J. Chem. Phys. 39, 1397 (1963). Hoffmann, R.; Lipscomb, W. N. ibid. 36, 2179 (1962). 37, 2872 (1982). The extended Huckel method was devised in the exciting atmosphere of the Lipscomb laboratory; L. L. Lohr played an essential part in its formulation. The first Cornell implementation to an extended material was made by M.-H. Whangbo [Whangbo, M.-H.; Hoffmann, R. J. Am. Chem. Soc. 100, 6093 (1978). Whangbo, M.-H.; Hoffmann, R.; Woodward, R. B. Proc. Roy. Soc. A366, 23 (1979).

5. (a) Burdett, J. K. Nature 279, 121 (1979).
   (b) Burdett, J. K. in Ref. 1d.
   (c) Burdett, J. K. Accts. Chem. Res. 15, 34 (1982) and references therein.
   (d) Burdett, J. K. Accts. Chem. Res. 21, 189 (1988).

6. Burdett, J. K. Progr. Sol. St. Chem. 15, 173 (1984).

7. Whangbo, M.-H. in "Extended Linear Chain Compounds," Miller, J. S. (ed.), Plenum Press, New York, 1982, p. 127.

8. Whangbo, M.-H. in "Crystal Chemistry and Properties of Materials with Quasi-One Dimensional Structures," Rouxel, J. (ed.), Reidel, Dordrecht, 1986, p. 27.

9. The modern classic solid state physics text is by Ashcroft, N. W. and Mermin, N. D. "Solid State Physics," Holt, Rinehart and Winston, New York, 1976. Three other introductions to the field, ones that I think are pedagogically effective and accessible to chemists, are: Harrison, W. A. "Solid State Theory," Dover, New York, 1980. Harrison, W. A.: "Electronic Structure and the Properties of Solids," W. H. Freeman, San Francisco, 1980; Altman, S. L. "Band Theory of Metals," Pergamon Press, New York, 1970.

10. Excellent introductions to this subject, written in the spirit of this book, and with a chemical audience in mind, are to be found in Refs. 6 and 8, in the book by P. A. Cox in Ref. 1, and in the article by Kertesz, M. Int. Rev. Phys. Chem. 4, 125 (1985).

11. Albright, T. A.; Burdett, J. K., Whangbo, M. H. "Orbital Interactions in Chemistry," Wiley-Interscience, New York, 1985, Chap. 20.

12. For a review of this fascinating class of materials see Williams, J. R. Adv. Inorg. Chem. Radiochem. 26, 235 (1983). Related to these molecules are the "platinum blues", in which one finds oxidized Pt(II) oligomers. For a leading reference, see: O'Halloran, T.

V.; Mascharak, P. K.; Williams, I. D.; Roberts, M. M.; Lippard, S. J. *Inorg. Chem. 26,* 1261 (1987).

13. Cotton, F. A.; Walton, R. A. "Multiple Bonds Between Metal Atoms," John Wiley and Sons, New York, 1982 and references therein.

14. For more information on the platinocyanides, see Refs. 7, 8, and 12.

15. For this task, and for band structure calculation in general, a chemist needs to learn group theory in the solid state. For a lucid introduction, see Tinkham, M. J.: "Group Theory and Quantum Mechanics," McGraw-Hill, New York, 1964; Burns, G.; Glazer, A. M. "Space Groups for Solid State Scientists," Academic Press, New York, 1978; Madelung, O. "Introduction to Solid State Theory," Springer Verlag, Berlin, 1978.

16. Tremel, W.; Hoffmann, R. *J. Am. Chem. Soc. 109,* 124 (1987); *Inorg. Chem. 26,* 118 (1987); Keszler, D.; Hoffmann, R. *J. Am. Chem. Soc., 109,* 118 (1987).

17. Zonnevylle, M. C.; Hoffmann, R. *Langmuir 3,* 452 (1987).

18. Trinquier, G.; Hoffmann, R. *J. Phys. Chem. 88,* 6696 (1984).

19. Li, J.; Hoffmann, R. *Z. Naturforsch 41b,* 1399 (1986).

20. The following references are just ways in to the vast literature of reconstruction: Kleinle, G.; Penka, V.; Behm, R. J.; Ertl, G.; Moritz, W. *Phys. Rev. Lett., 58,* 148 (1987); Daum, W.; Lehwald, S.; Ibach, H. *Surf. Sci. 178,* 528 (1986); Van Hove, M. A.; Koestner, R. J.; Stair, P. C.; Biberian, J. P.; Kesmodel, L. L.; Bartos, I.; Somorjai, G. A. *Surf. Sci. 103,* 189 (1981); Inglesfield, J. E.; *Progr. Surf. Sci. 20,* 105 (1985); King, D. A. *Phys. Scripta 4,* 34 (1983); Chan, C.-M.; Van Hove, M. A. *Surf. Sci. 171,* 226 (1986); Christmann, K. *Z. Phys. Chem. 154,* 145 (1987).

21. (a) Wade, K. *Chem. Commun.* 792 (1971); *Inorg. Nucl. Lett., 8,* 559 (1972); "Electron Deficient Compounds," Nelson, London, 1971.
    (b) Mingos, D. M. P. *Nature 236,* 99 (1972).

22. It should be clear that this is just one possible methodology. Many other theoretical chemists and physicists prefer a cluster approach, i.e., a finite group of metal atoms and single adsorbate. The local chemical action is then emphasized, though one has to worry about how the cluster is terminated, or what the end effects are. More recently, people have begun to consider embeddings of clusters in extended structures.

23. For more on the electronic structure of rutile and related compounds, see Burdett, J. K.; Hughbanks, T. *Inorg. Chem. 24,* 1741 (1985); Burdett, J. K. *Inorg. Chem. 24,* 2244 (1985); Mattheiss, L. F. *Phys. Rev.,* B13, 2433 (1976).

24. Some leading references to early band structure calculations are the following:
    (a) Mattheiss, L. F. *Phys. Rev. Lett. 58,* 1028 (1987).
    (b) Yu, J. J.; Freeman, A. J.; Xu, J.-H. *Phys. Rev. Lett., 58,* 1035 (1987); Massidda, S.; Yu, J.; Freeman, A. J.; Koelling, D. D. *Phys. Lett.* A122, 198 (1987).
    (c) Whangbo, M.-H.; Evain, M.; Beno, M. A.; Williams, J. M. *Inorg. Chem. 26,* 1829 (1987); *ibid., 26,* 1831 (1987); Whangbo, M.-H.; Evain, M.; Beno, M. A.; Geiser, U.; Williams, J. M., *ibid., 26,* 2566 (1987).
    (d) Fujiwara, R.; Hatsugai, Y. *Jap. J. Appl. Phys. 26,* L716 (1987).

25. Goodenough, J. B. "Magnetism and the Chemical Bond," Krieger, New York, 1976.

26. Mulliken, R. S. *J. Chem. Phys. 23,* 1833, 2343 (1955).

27. For further details see Sung, S.; Hoffmann, R. *J. Am. Chem Soc. 107,* 578 (1985).

28. (a) For discussion of other calculations of CO on Ni surfaces, see Kasowski, R. V.; Rhodin, T.; Tsai, M.-H. *Appl. Phys.* A41, 61 (1986) and references therein.
    (b) Avouris, Ph.; Bagus, P. S.; Nelin, C. J. *J. Electron. Spectr. Rel. Phen.* 38, 269 (1986) stressed the importance of this phenomenon. We disagree on its magnitude, in that we find $p_{x,y}$ mixing into the main $2\pi^*$ density at -7 eV small. The COOP curve, to be shown later in Fig. 24, indicates that the density in this peak is Ni–C antibonding.

29. For further details, see Silvestre, J.; Hoffmann, R. *Langmuir 1,* 621 (1985).

30. See Saillard, J.-Y.; Hoffmann, R. *J. Am. Chem. Soc. 106*, 2006 (1984) for further details.
31. (a) Shustorovich, E. *J. Phys. Chem. 87*, 14 (1983).
    (b) Shustorovich, E. *Surf. Sci. Rep. 6*, 1 (1986).
32. Shustorovich, E.; Baetzold, R. C. *Science 227*, 876 (1985).
33. There is actually some disagreement in the literature on the relative role of C–H and H–H σ and σ* levels in interactions with metal surfaces. A. B. Anderson finds σ donation playing the major role. See, for example, Anderson, A. B. *J. Am. Chem Soc. 99*, 696 (1977) and subsequent papers.
34. COOP was introduced for extended systems in papers by
    (a) Hughbanks, R.; Hoffmann, R. *J. Am. Chem. Soc. 105*, 3528 (1983).
    (b) Wijeyesekera, S. D.; Hoffmann, R. *Organometal. 3*, 949 (1984).
    (c) Kertesz, M.; Hoffmann, R. *J. Am. Chem Soc. 106*, 3453 (1984).
    (d) An analogous index in the Hückel model, a bond order density, was introduced earlier by van Doorn, W.; Koutecký, J. *Int. J. Quantum Chem. 12*, Suppl. 2, 13 (1977).
35. This is from an extended Hückel calculation (Ref. 30). For better estimates, see Ref. 40.
36. For some references to this story, see note 5 in Ref. 29. An important early LEED analysis here was by Kesmodel, L. L.; Dubois, L. H.; Somorjai, G. A. *J. Chem. Phys. 70*, 2180 (1979).
37. Pearson, W. B. *J. Sol. State Chem. 56*, 278 (1985) and references therein.
38. Mewis, A. *Z. Naturforsch. 35b*, 141 (1980).
39. (a) Zheng, C.; Hoffmann, R. *J. Phys. Chem. 89*, 4175 (1985).
    (b) Zheng, C.; Hoffmann, R. *Z. Naturforsch. 41b*, 292 (1986).
    (c) Zheng, C.; Hoffmann, R. *J. Am. Chem. Soc. 108*, 3078 (1986).
40. Andersen, O. K. in "The Electronic Structure of Complex Systems," Phariseau, P.; Temmerman, W. M. (ed.), Plenum Press, New York, 1984; Andersen, O. K. in "Highlights of Condensed Matter Physics," Bassani, F.; Fumi, F.; Tossi, M. P. (eds.), North-Holland, New York, 1985. See also Varma, C. M.; Wilson, A. J. *Phys. Rev. B22*, 3795 (1980).
41. (a) The frontier orbital concept is a torrent into which flowed many streams. The ideas of Fukui were a crucial early contribution (the relevant papers are cited by Fukui, K. *Science 218*, 747 (1982) as was the perturbation theory based PMO approach of Dewar (see Dewar, M. J. S. "The Molecular Orbital Theory of Organic Chemistry," McGraw-Hill, New York, 1969 for the original references). The work of Salem was important (see Jorgensen, W. L., Salem, L. "The Organic Chemist's Book of Orbitals," Academic Press, New York, 1973, for references and a model portrayal, in the discussion preceding the drawings, of the way of thinking that my coworkers and I also espoused). The Albright, Burdett, and Whangbo text (Ref. 11) carries through this philosophy for inorganic systems and is also an excellent source of references.
    (b) For surfaces, the frontier orbital approach is really there in the pioneering work of Blyholder, G., *J. Phys. Chem. 68*, 2772 (1964). In our work in the surface field, we first used this way of thinking in Ref. 30, a side-by-side analysis of molecular and surface H–H and C–H activation.
42. Murrell, J. N.; Randić, M.; Williams, D. R. *Proc. Roy. Soc. A284*, 566 (1965); Devaquet, A.; Salem, L. *J. Am. Chem. Soc. 91*, 379 (1969); Fukui, K.; Fujimoto, H. *Bull Chem. Soc. Jpn. 41*, 1984 (1968); Baba, M.; Suzuki, S.; Takemura, T. *J. Chem. Phys. 50*, 2078 (1969); Whangbo, M.-H.; Wolfe, S. *Can. J. Chem. 54*, 949 (1976); Morokuma, K. *J. Chem. Phys. 55*, 1236 (1971) is a sampling of these.
43. See, for instance, Kang, A. B.; Anderson, A. B. *Surf. Sci. 155*, 639 (1985).
44. (a) Gadzuk, J. W., *Surf Sci. 43*, 44 (1974).
    (b) See also Varma, C. M.; Wilson, A. J. *Phys. Rev. B. 22*, 3795 (1980); Wilson, A. J.; Varma, C. M. *Phys. Rev. B. 22*, 3805 (1980); Andreoni, W.; Varma, C. M. *Phys. Rev. B. 23*, 437 (1981).
45. Grimley, T. B. *J. Vac. Sci Technol. 8*, 31 (1971); in Ricca, F. (ed.). "Adsorption-

Desorption Phenomena,'' Academic Press, New York, 1972, p. 215 and subsequent papers. See also Thorpe, B. J. *Surf. Sci. 33*, 306 (1972).

46. (a) van Santen, R. A. Proc. 8th Congr. Catal., Springer-Verlag, Berlin, 1984, Vol. 4, p. 97; *J. Chem. Soc. Far. Trans. 83*, 1915 (1987).
    (b) LaFemina, J. P.; Lowe, J. P. *J. Am. Chem. Soc. 108*, 2527 (1986).
    (c) Fujimoto, M. *Accts. Chem. Res. 20*, 448 (1987); *J. Phys. Chem. 41*, 3555 (1987).

47. Salem, L.; Leforestier, C. *Surf. Sci. 82*, 390 (1979); Salem, L.; Elliot, R. *J. Mol. Struct. Theochem. 93*, 75 (1983); Salem, L. *J. Phys. Chem 89*, 5576 (1985); Salem, L.; Lefebvre, R. *Chem. Phys. Lett. 122*, 342 (1985).

48. (a) Banholzer, W. F.; Park, Y. O.; Mak, K. M.; Masel, R. I. *Surf. Sci. 128*, 176 (1983); Masel, R. I., to be published.

49. See Elian, M.; Hoffmann, R. *Inorg. Chem. 14*, 1058 (1975).

50. The effect mentioned here has also been noted by Raatz, R.; Salahub, D. R. *Surf. Sci. 146*, L609 (1984); Salahub, D. R.; Raatz, F. *Int. J. Quantum Chem. Symp. No. 18*, 173 (1984); Andzelm, J.; Salahub, D. R. *Int. J. Quantum Chem. 29*, 1091 (1986).

51. Dewar, M. J. S. *Bull Soc. Chim. Fr. 18*, C71 (1951); Chatt, J.; Duncanson, L. A. *J. Chem. Soc.* 2939 (1953).

52. (a) Muetterties, E. L. *Chem. Soc. Revs. 11*, 283 (1982); *Angew. Chem. Int. Ed. Engl. 17*, 545 (1978); Muetterties, E. L.; Rhodin, T. N. *Chem. Revs. 79*, 91 (1979).
    Albert, M. R.; Yates, J. T. Jr. "The Surface Scientist's Guide to Organometallic Chemistry," American Chemical Society, Washington, 1987.

53. Garfunkel, E. L.; Minot, C.; Gavezzotti, A.; Simonetta, M. *Surf. Sci. 167*, 177 (1986); Garfunkel, E. L.; Feng, X. *Surf. Sci. 176*, 445 (1986); Minot, C.; Bigot, B.; Hariti, A. *Nouv. J. Chim. 10*, 461 (1986). See also Ref. 50 and Shustorovich, E. M. *Surf. Sci. 150*, L115 (1985).

54. Tang, S. L.; Lee, M. B.; Beckerle, J. D.; Hines, M. A.; Ceyer, S. T. *J. Chem. Phys. 82*, 2826 (1985); Tang, S. L.; Beckerle, J. D.; Lee, M. B.; Ceyer, S. T. *ibid., 84*, 6488 (1986). See also Lo, T.-C.; Ehrlich, B. *Surf. Sci. 179*, L19 (1987); Steinruck, H.-P.; Hamza, A. V.; Madix, R. J. *Surf. Sci. 173*, L571 (1986); Hamza, A. V.; Steinruck, H.-P.; Madix, R. J. *J. Chem. Phys. 86*, 6506 (1987) and references in these.

55. Several such minima have been computed in the interaction of atomic and molecular adsorbates with Ni clusters: P. E. M. Siegbahn, private communication.

56. (a) Kang and Anderson in Ref. 43 and papers cited there.
    Harris, J.; Andersson, S. *Phys. Rev. Lett. 55*, 1583 (1985).

57. Mango, F. D.; Schachtschneider, J. H. *J. Am. Chem. Soc. 89*, 2848 (1967); Mango, F. D. *Coord. Chem. Revs. 15*, 109 (1978).

58. Hoffmann, R. in "IUPAC: Frontiers of Chemistry," Laidler, K. J. (ed.), Pergamon Press, Oxford, 1982; Tatsumi, K.; Hoffmann, R.; Yamamoto, A.; Stille, J. K. *Bull. Chem. Soc. Jpn. 54*, 1857 (1981) and references therein.

59. See Ref. 29 for detailed discussion of this phenomenon.

60. Anderson, A. B.; Mehandru, S. P. *Surf. Sci. 136*, 398 (1984); Kang, D. B.; Anderson, A. B. *ibid. 165*, 221 (1986).

61. For a review, see Chevrel, R. in "Superconductor Materials Science: Metallurgy, Fabrication, and Applications," Foner, S.; Schwartz, B. B. (eds.), Plenum Press, New York, 1981, Chap. 10.

62. Lin, J.-H.; Burdett, J. K. *Inorg. Chem. 5*, 21 (1982).

63. (a) Hughbanks, T.; Hoffmann, R. *J. Am. Chem. Soc. 105*, 1150 (1983).
    (b) Such a synthesis, by a different route, has recently been achieved: Saito, T.: Yamamoto, N.: Yamagata, T.; Imoto, M. *J. Am. Chem. Soc. 110*, 1646 (1988).

64. See also Andersen, O. K.; Klose, W.; Nohl, H. *Phys. Rev. B17*, 1760 (1977); Nohl, H.; Klose, W.; Andersen, O. K. in "Superconductivity in Ternary Compounds," Fisher,

∅.; Maple, M. B. (eds.), Springer-Verlag, New York, 1981, Chap. 6; Bullett, D. W. *Phys. Rev. Lett. 44*, 178 (1980).

65. See. Ref. 39c and papers cited therein.

66. This drawing is taken from Ref. 11. Excellent descriptions of the folding-back process and its importance are to be found in Refs. 6 and 8.

67. The original reference is Jahn, H. A.; Teller, E. *Proc. Roy. Soc. A161*, 220 (1937). A discussion of the utility of this theorem in deriving molecular geometries is given by Burdett, J. K. "Molecular Shapes," John Wiley, New York, 1960; Burdett, J. K. *Inorg. Chem. 20*, 1959 (1981) and references cited therein. See also Refs. 6, 8 and Pearson, R. G. "Symmetry Rules for Chemical Reactions," John Wiley and Sons, New York, 1976. The first application of the Jahn–Teller argument to stereochemical problems and, incidentally, to condensed phases was made by Dunitz, J. D.; Orgel, L. E. *J. Phys. Chem. Sol. 20*, 318 (1957); *Adv. Inorg. Chem. Radiochem. 2*, 1 (1960); Orgel, L. E.; Dunitz, J. D. *Nature 179*, 462 (1957).

68. Peierls, R. E. "Quantum Theory of Solids," Oxford University Press, Oxford, 1972.

69. Longuet-Higgins, H. C.; Salem, L. *Proc. Roy. Soc. A251*, 172 (1959).

70. For leading references, see the review by Kertesz, M. *Adv. Quant. Chem. 15*, 161 (1982).

71. Pearson, W. B. *Z. Kristallogr. 171*, 23 (1985) and references therein.

72. Hulliger, F.; Schmelczer, R.; Schwarzenbach, D. *J. Sol. State Chem. 21*, 371 (1977).

73. Schmelczer, R.; Schwarzenbach, D.; Hulliger, F. *Z. Naturforsch. 36b*, 463 (1981) and references therein.

74. (a) Burdett, J. K.; Haaland, P.; McLarnan, T. J. *J. Chem. Phys. 75*, 5774 (1981).
    (b) Burdett, J. K.; Lee, S. *J. Am. Chem. Soc. 105*, 1079 (1983).
    (c) Burdett, J. K.; McLarnan, T. J. *J. Chem. Phys. 75*, 5764 (1981).
    (d) Littlewood, P. B. *CRC Critical Reviews in Sol. State and Mat. Sci. 11*, 229 (1984).

75. Tremel, W.; Hoffmann, R. *J. Am. Chem. Soc. 108*, 5174 (1986) and references cited therein; Silvestre, J.; Tremel, W.; Hoffmann, R. *J. Less Common Met. 116*, 113 (1986).

76. Franzen, H. F.; Burger, T. J. *J. Chem. Phys. 49*, 2268 (1968); Franzen, H. F.; Wiegers, G. A. *J. Sol. State Chem., 13*, 114 (1975).

77. See Ref. 75 and Kertesz, M.; Hoffmann, R. *J. Am. Chem. Soc. 106*, 3453 (1984).

78. Zheng, C.; Apeloig, Y.; Hoffmann, R. *J. Am. Chem. Soc. 110*, 749 (1988).

79. Brodén, G.; Rhodin, T. N.; Brucker, C.; Benbow, H.; Hurych, Z. *Surf. Sci. 59*, 593 (1976); Engel, T.; Ertl, G. *Adv. Catal. 28*, 1 (1979).

80. Shinn, N. D.; Madey, R. E. *Phys. Rev. Lett. 53*, 2481 (1984); *J. Chem. Phys. 83*, 5928 (1985); *Phys. Rev. B33*, 1464 (1986); Benndorf, C.; Kruger, B.; Thieme, F. *Surf. Sci. 163*, L675 (1985). On Pt_3Ti a similar assignment has been made: Bardi, U.; Dahlgren, D.; Ross, H. *J. Catal. 100*, 196 (1986). Also for CN on Cu and Pd surfaces: Kordesch, M. E.; Stenzel, W.; Conrad, H. *J. Electr. Spectr. Rel. Phen. 38*, 89 (1986); Somers, J.; Kordesch, M. E.; Lindner, Th.; Conrad, H.; Bradshaw, A. M.; Williams, G. P. *Surf. Sci. 188*, L693 (1987).

81. Mehandru, S. P.; Anderson, A. B. *Surf. Sci. 169*, L281 (1986); Mehandru, S. P.; Anderson, A. B.; Ross, P. N. *J. Catal. 100*, 210 (1986).

82. For a recent review, see Anderson, R. B. "The Fischer–Tropsch Synthesis," Academic Press, New York, 1984.

83. (a) See, however, Baetzold, R. *J. Am. Chem. Soc. 105*, 4271 (1983) and Wittrig, T. S.; Szoromi, P. D.; Weinberg, W. M. *J. Chem. Phys. 76*, 3305 (1982).
    (b) See also Lichtenberger, D. L.; Kellogg, G. E. *Accounts Chem. Res. 20*, 379 (1979) and Wilker, C. N.; Hoffmann, R.; Eisenstein, O. *Nouv. J. Chim. 7*, 535 (1983).
    (c) Siegbahn, P. E. M.; Blomberg, M. R. A.; Bauschlicher, C. W. Jr. *J. Chem. Phys. 81*, 2103 (1984); Upton, T. H. *J. Am. Chem. Soc. 106*, 1561 (1984).
    (d) Feibelman, P. J.; Hamann, D. R. *Surf. Sci. 149*, 48 (1985).

84. Reviews of some of the theoretical work on surfaces may be found in Gavezzotti, A.; Simonetta, M. *Adv. Quantum Chem. 12*, 103 (1980); Messmer, R. P. in "Semiempirical Methods in Electronic Structure Calculations, Part B: Applications," Segal, G. A. (ed.), Plenum Press, New York, 1977; Koutecky, J.; Fantucci, P. *Chem. Revs., 86*, 539 (1986); Cohen, M. L. *Ann. Rev. Phys. Chem. 34*, 537 (1984).

85. Lundqvist, B. I. *Chemica Scripta 26*, 423 (1986); *Vacuum, 33*, 639 (1983); in "Many-Body Phenomena at Surfaces," Langreth, D. C.; Suhl, H. (eds.), Academic Press, New York, 1984, p. 93; Lang, N. D.; Williams, A. R. *Phys. Rev. Lett. 37*, 212 (1976); *Phys. Rev. B18*, 616 (1978); Lang, N. D.; Nørskov, J. K. in "Proceedings 8th International Congress on Catalysis," Berlin, 1984, Vol. 4, p. 85; Nørskov, J. K.; Lang, N. D. *Phys. Rev., B21*, 2136 (1980); Lang, N. D. in "Theory of the Inhomogeneous Electron Gas," Lundqvist, S.; March, N. H. (eds.), Plenum Press, New York, 1983; Hjelmberg, H.; Lundqvist, B. I.; Nørskov, J. K. *Physica Scripta 20*, 192 (1979).

86. Kohn, W.; Sham, L. J. *Phys. Rev. 140*, A1133 (1965).

87. Nørskov, J. K.; Houmøller, A.; Johansson, P. K.; Lundqvist, B. I. *Phys. Rev. Lett. 46*, 257 (1981); Lundqvist, B. I.; Nørskov, J. K.; Hjelmberg, H. *Surf. Sci. 80*, 441 (1979); Nørskov, J. K.; Holloway, S.; Lang, N. D. *Surf. Sci. 137*, 65 (1984).

88. Figure 18 is for a perpendicular H_2 approach. We should really compare a parallel one to the Mg case, and the features of that parallel approach, while not given here, resemble Fig. 18.

89. See, for instance, Refs. 43, 60, and 81.

90. Shustorovich, E. *Surf. Sci. Rep. 6*, 1 (1986) and references therein; Shustorovich, E.; Baetzold, R. C. *Science 227*, 876 (1985); Baetzold, R. in "Catalysis," Moffat, J. (ed.), Hutchinson-Ross, 1988.

91. Minot, C.; Bigot, B.; Hariti, A. *Nouv. J. Chim. 10*, 461 (1986); *J. Am. Chem. Soc. 108*, 196 (1986); Minot, C.; Van Hove, M. A.; Somorjai, G. A. *Surf. Sci. 127*, 441 (1983); Bigot, B.; Minot, C. *J. Am. Chem. Soc. 106*, 6601 (1984) and subsequent references.

92. For references, see Wijeyesekera, S. D.; Hoffmann, R. *Inor. Chem. 22*, 3287 (1983).

93. Bronger, W. *Angew. Chem. 93*, 12 (1981); *Angew. Chem. Int. Ed. Engl. 20*, 52 (1981).

94. Silvestre, J.; Hoffmann, R. *Inorg. Chem. 24*, 4108 (1985).

Index